World Tourism Cities

This book presents new research on the capacity of big cities to generate new tourism areas as visitors discover and help create new urban experiences off the beaten track. It examines similarities and differences in these processes in a group of established world cities located in the global circuits of tourism. The cities featured are Berlin, New York, London, Paris, and Sydney.

In these cities experienced city visitors are contributing to the 'discovery' of new places to visit. Many neighbourhoods close to the historic centre and to traditional attractions offer the mix of cultural difference and consumption opportunities that can create new experiences for distinctive groups of city users. Each of the cities included in the book offers rich experiences of the re-imagining and re-branding of neighbourhoods off the beaten track, informative stories of the complex relationships between visitors, residents and others and of the ambitions of public policy to reproduce these new tourism experiences in other parts of the city. *World Tourism Cities* brings together current research in each of the cities and relates the often separate field of tourism research to some of the mainstream themes of debate in urban studies addressing topics such as consumption, markets and spaces.

Drawing on original research in this important group of cities this book has significant messages for public policy. In addition, the book engages directly with a range of important current academic debates: about world cities, about cities as sites of consumption and about the smaller scales at which urban neighbourhoods are being transformed. The range of cities and the messages about the making of attractive places provides a timely resource for those focused in this area and the book will also have an appeal among those experienced and sophisticated city users that it focuses on.

Robert Maitland is Reader in Tourism at the University of Westminster, London. His research focuses on tourism in cities, particularly world cities and national capitals, and on tourism policy. He has led funded research projects, written articles and books and advised government on these themes. Current research examines visitors' role in the creation of new tourist areas in London, and tourism in national capitals.

Peter Newman is Professor of Comparative Urban Planning at the School of Architecture and the Built Environment, University of Westminster, London. He has written widely on European cities, governance and planning. His most recent book is a study of urban planning and city management issues in 'world cities' in North America, Asia and Europe (*Planning World Cities* Palgrave 2005).

Contemporary geographies of leisure, tourism and mobility
Series edited by C. Michael Hall
Professor at the Department of Management, College of Business & Economics, University of Canterbury, Private Bag 4800, Christchurch, New Zealand

The aim of this series is to explore and communicate the intersections and relationships between leisure, tourism and human mobility within the social sciences. It will incorporate both traditional and new perspectives on leisure and tourism from contemporary geography, for example notions of identity and representation and culture, while also providing for perspectives from cognate areas such as anthropology, cultural studies, gastronomy and food studies, marketing, policy studies and political economy, regional and urban planning, and sociology within the development of an integrated field of leisure and tourism studies.

Increasingly, tourism and leisure are regarded as steps in a continuum of human mobility. Inclusion of mobility in the series offers the prospect to examine the relationship between tourism and migration, the sojourner, educational travel, and second home and retirement travel phenomena.

The series comprises two strands:

CONTEMPORARY GEOGRAPHIES OF LEISURE, TOURISM AND MOBILITY

This aims to address the needs of students and academics, and the titles will be published in hardback and paperback. Titles include:

The Moralisation of Tourism
Sun, sand . . . and saving the world?
J. Butcher

The Ethics of Tourism Development
M. Smith and R. Duffy

Tourism in the Caribbean
Trends, development, prospects
Edited by D. T. Duval

Qualitative Research in Tourism
Ontologies, epistemologies and methodologies
Edited by J. Phillimore and L. Goodson

The Media and the Tourist Imagination
Converging cultures
Edited by D. Crouch, R. Jackson and F. Thompson

Tourism and Global Environmental Change
Ecological, social, economic and political interrelationships
Edited by S. Gössling and C. M. Hall

Forthcoming:

Understanding and Managing Tourism Impacts
M. Hall and A. Lew

ROUTLEDGE STUDIES IN CONTEMPORARY GEOGRAPHIES OF LEISURE, TOURISM AND MOBILITY

This is a forum for innovative new research intended for research students and academics, and the titles will be available in hardback only. Titles include:

Living with Tourism
Negotiating identities in a Turkish village
H. Tucker

Tourism, Diasporas and Space
T. Coles and D. J. Timothy

Tourism and Postcolonialism
Contested discourses, identities and representations
C. M. Hall and H. Tucker

Tourism, Religion and Spiritual Journeys
D. J. Timothy and D. H. Olsen

China's Outbound Tourism
W. G. Arlt

Tourism, Power and Space
A. Church and T. Coles

Tourism, Ethnic Diversity and the City
J. Rath

Ecotourism, NGOs and Development
A critical analysis
J. Butcher

Tourism and the Consumption of Wildlife
Hunting, shooting and sport fishing
B. Lovelock

Tourism, Creativity and Development
G. Richards and J. Wilson

Tourism at the Grassroots
J. Connell and B. Rugendyke

Tourism and Innovation
M. Hall and A. Williams

World Tourism Cities
Developing tourism off the beaten track
R. Maitland and P. Newman

Forthcoming:

Understanding and Managing Tourism Impacts
M. Hall and A. Lew

Tourism Geography
A new synthesis
S. Williams

Tourism, Performance and the Everyday
Consuming the Orient
M. Haldrup and J. Larsen

World Tourism Cities
Developing tourism off the beaten track

**Edited by Robert Maitland and
Peter Newman**

Routledge
Taylor & Francis Group

LONDON AND NEW YORK

First published 2009 by Routledge
2 Park Square, Milton Park, Abingdon, Oxon, OX14 4RN

Simultaneously published in the USA and Canada
by Routledge
270 Madison Avenue, New York, NY 10016

*Routledge is an imprint of the Taylor & Francis Group,
an informa business*

© 2009 Editorial matter by Robert Maitland and Peter Newman and
selections by contributors

Typeset in Times NR MT by Graphicraft Limited, Hong Kong

British Library Cataloguing in Publication Data
A catalogue record for this book is available from the
British Library

Library of Congress Cataloguing in Publication Data
World tourism cities : developing tourism off the beaten track /
edited by Robert Maitland and Peter Newman.
p. cm. — (Routledge contemporary geographies of leisure, tourism
and mobility ; 10)
Includes bibliographical references.
1. Tourism. 2. Tourism—Research—Methodology. I. Maitland,
Robert. II. Newman, Peter, 1949–
G155.A1W685 2008
910.68—dc22
2008023510

ISBN 13: 978–0–415–45198–7 (hbk)
ISBN 13: 978–0–203–88656–4 (ebk)

ISBN 10: 0–415–45198–1 (hbk)
ISBN 10: 0–203–88656–9 (ebk)

For Jane, for being there.

Contents

List of tables and figures

TABLES

FIGURES

Acknowledgements

The idea for this book grew out of discussions at conferences and elsewhere with a number of colleagues, including those who have contributed chapters. In addition to them we would also like to thank Terry Nichols Clark and Lily Hoffman; particular thanks to Ilaria Pappalepore, doctoral candidate in the School of Architecture and Built Environment, University of Westminster for her help in assembling data.

Robert Maitland and Peter Newman

Jill Gross would like to thank Roberto Genoves, Mark Hogan, Phillip Mallory, Megan Murphy and Jose Roman – students participating in an applied urban research workshop at the Hunter College Graduate Program in Urban Affairs and Planning.

Johannes Novy and Sandra Huning would like to thank their interview partners in Berlin (especially Kathrin Klisch, Natascha Kompatzki and Ursula Luchner-Bruck). They would also like to express gratitude to Uwe-Jens Walther, Susan Fainstein, Peter Marcuse, and Karsten Foth as well as several other staff members and students at the Institute for Sociology and the Center for Metropolitan Studies (CMS) of Berlin's Technical University and Columbia University's Graduate School for Architecture, Planning and Preservation.

Contributors

Tony Griffin is a Senior Lecturer in the School of Leisure, Sport and Tourism at the University of Technology, Sydney. With a professional background in urban planning, he has published extensively on subjects ranging from hotel development to sustainable tourism. His recent research has focused on understanding visitor experiences in a variety of contexts including urban tourism precincts, national parks and wine tourism.

Jill Simone Gross is Associate Professor of Political Science and Director of the Masters Program in Urban Affairs at Hunter College of the City University of New York. She is co-editor of *Governing Cities in a Global Era*. Palgrave 2008.

Bruce Hayllar PhD is Associate Professor, and Head, School of Leisure, Sport and Tourism, University of Technology, Sydney. Bruce has a particular interest in the experience of people in learning and leisure environments and has applied his interest in phenomenology to inform this understanding. His most recent work has been a two-year national project, funded by the Sustainable Tourism Co-operative Research Centre, which examined the experience of tourists in urban precincts.

Dr Sandra Huning is a spatial planner with a focus on urban studies and planning theory. She was postdoc fellow at the Center for Metropolitan Studies and researcher at the German Academy of Sciences Leopoldina. Currently, she is working on the interrelations of global change and regional development at the Berlin-Brandenburg Academy of Sciences and Humanities.

Patrizia Ingallina is Professor at the Université des Sciences et Technologies de Lille, IAUL (Institut d'Aménagement d'Urbanisme de Lille). Her most recent publications include, *Le Projet Urbain*, Que-sais-je? Paris 2008 and 'City marketing et espaces de consommation. Les nouveaux enjeux de l'attractivité urbaine', *Urbanisme*, 344, 2005 (with JY Park).

Robert Maitland is Reader in Tourism at the University of Westminster, London. His research focuses on tourism in cities, particularly world cities and national capitals, and on tourism policy. He has led funded research projects, written articles and books and advised government on these themes. Current research examines visitors' role in the creation of new tourist areas in London, and tourism in national capitals.

Peter Newman is Professor of Comparative Urban Planning at the School of Architecture and the Built Environment, University of Westminster, London. He has written widely on European cities, governance and planning. His most recent book is a study of urban planning and city management issues in 'world cities' in North America, Asia and Europe (*Planning World Cities*. Palgrave 2005).

Johannes Novy is a PhD Candidate in Urban Planning at Columbia University's Graduate School for Architecture, Planning and Preservation. He is also a fellow at the Center of Metropolitan Studies (CMS) in Berlin. His research interests include planning history and theory, urban tourism, as well as urban development in North America and Europe.

Jungyoon Park is a doctoral candidate at the Institut d'Urbanisme de Paris, University of Paris 12.

1 Developing world tourism cities

Robert Maitland and Peter Newman

INTRODUCTION

Urban tourism has been an inseparable part of the transformation of many cities over the past few decades. The waterfront developments, repackaged culture and heritage and café culture that signalled a new direction for many northern cities have been echoed in east Asia and other global regions. Place wars, imagineering and mega-projects are equally widespread and contribute to new city images that both confirm the urban preferences of those residents and businesses that are winners in urban change, and impact on the choices of visitors.

Tourism has boomed. Whilst there are precise official definitions of tourists and tourism (World Tourism Organisation and United Nations 1994) the conceptual difficulties of distinguishing tourism from other activities in cities and from other forms of mobility have meant that some commentators have referred to 'the end of tourism' as a discrete activity (Urry 1995, p. 150). There has been such a rapid growth in mobilities of many sorts that:

> all the world seems to be on the move . . . there are over 700 million legal passenger arrivals each year (compared to 25 million in 1950) . . . the inter-net has [almost] 1 billion users world wide . . . new forms of virtual and imaginative travel are emerging [so that there are] diverse yet intersecting mobilities [that] have many consequences for different people and places.
> (Sheller and Urry 2006, p. 207)

Urban tourism is a vital part of this growth in mobility, yet it is increasingly difficult to see tourists as separated from other urban processes. In this book, we take up the challenge of understanding the interactions of visitors, residents, city workers and other city users in processes of urban change. We aim to explore systematically themes in tourism research about the discovery and creation of new urban places and the intersecting mobilities and urban preferences that may be shared between different groups. We relate discussion about the impacts of visitors and their concern with the quality and authenticity of urban experiences to the broader debates in urban studies about a middle class return to the city, about consumption and everyday life, and a new concern with quality of life and amenity in attracting and retaining residents and workers. These broad themes about class, social values, consumption, and creativity help account for change in many cities and have their visible impacts on public policy as city leaders chase a creative class, advertise quality of life attributes and seek the apparent benefits of a visitor economy. In many ways, such approaches have become typical of old industrial cities and the transformation of central European cities and also find their echoes in more established, historic tourist cities, and growing tourist cities in east Asia. These are important processes in the cities we examine in this book. However, we argue that some of these processes are pronounced in the group of cities that we term world tourism cities (WTC). These are large polycentric cities offering a range of experiences and, as visitors move between and around established centres, they offer apparently seamless opportunities for adding new desirable places to explore to already crowded and diverse tourism possibilities. They are multi-functional cities well located in global circuits of both money and people. The idea of the world tourism city contrasts, however, with monocentric, less diverse but equally well known and well connected cities such as Venice or Las Vegas. Our idea of world tourism cities – we examine five in this book, London, Sydney, New York, Paris and Berlin – includes cities with substantial historical assets and iconic buildings, that are also centres of cultural excellence and, arising from their roles in global business networks, generate large numbers of business visitors in addition to those tourists attracted by tradition and cultural images. This group of cities is relatively rich, polycentric, multi-functional, culturally diverse and enjoys large flows of visitors. These are 'post colonial' cities, accommodating flows of migrants and where diversity can also be a marketing asset.

The characteristics of world tourism cities have another important consequence. The multiple social, physical and economic assets and polycentric spatial structures open up opportunities for tourism to develop away from traditional attractions and for visitors to discover new attractions off the beaten track. We argue that it is through the interactions of their multiple assets that world tourism cities have a considerable advantage over other cities in this ability to *produce* attractive places. Experienced urban tourists have an important role in enhancing the assets of world tourism cities. We suggest that it is through the work of visitors, residents and workers that these cities *produce* new localities and add to their already substantial advantage in the competition for reputation and visitors. These interactions between city users – visitors, residents and others – form an important dimension of our analysis in this book. Of course, the attractions of recently 'discovered' neighbourhoods can fade, and we expect experienced urban visitors to move on and continue to define new desirable locations within these large polycentric cities. Routes off the beaten track can become well trodden and may also change their character with the daily rhythms of the city. For example, a characterful street with charming local restaurants may seem too far off the beaten track as it transforms into a local drinking venue in the evening. Visitors and residents alike can become concerned at the loss of local distinctiveness. Two recent newspaper headlines, for example, bemoan the threat of the wrong sort of visitors to one well established (Champs Elysées) and another developing (Rue des Rosiers) tourist street – 'Sex, crime and brand names overwhelm the Elysian fields', 'For Paris's Jewish quarter, a fight to save its soul'.[1] Our expectations, preferences and use of the cities change. Some cities are better placed to respond to change and this capacity for change marks out our group of world tourism cities.

The chapters in this book explore a range of different experiences in a group of world tourism cities. These experiences expose the interaction of visitors with other groups in the, often complex, ways in which cities produce new urban experiences. Our aim in this chapter is to set the scene for those city studies. We outline the overall approach taken in the chapters and begin by attempting to show the links between themes in the wide range of work in tourism research and urban studies. Firstly, we examine changing understandings of the urban tourist. We argue that visitors have a vital role in making new urban experiences. We then develop the idea of the world tourism city as offering particular potential for interactions between groups of city users and then draw some linking themes from debates about what visitors want from cities and how residents express their urban preferences. Finally, we outline how these themes are followed through in the five cities.

CHANGING VIEWS OF THE TOURIST

The rapid growth of mobilities that we discussed at the start of the chapter gives the background to attempts to understand contemporary urban tourism. There is wide agreement that tourism has been affected by global economic restructuring that has changed the use of time and space (see discussions in for example Harvey 1989 and Meethan 2001). Clear demarcations between leisure and work places, leisure and work activities, and leisure and work time are being eroded, and with them the delineation between host and visitors, touristic and non-touristic activities. We should not exaggerate. In some cases, it remains possible to make clear, or at least workable, distinctions between hosts and visitors, places of work and residence and places of play to which people travel as tourists, and between touristic and non-touristic behaviour. Torremolinos in Spain is still primarily a holiday resort; visitors to The Gambia or Bhutan are unlikely to be confused with locals. However, in many destinations distinctions have become hazy and tourism cannot be seen as a separate activity confined to particular areas or to particular times (Franklin and Crang 2001). It is more pervasive and divisions between tourism and everyday life in the city have blurred. As spending on leisure rises, some cities become centres for entertainment and cultural consumption. At the same time, they are subject to rapid rises in tourism and non-tourism mobilities, the focus of multiplying and diverse images and are undergoing rapid economic and social change, with associated spatial reorganisation. Our argument is that these processes may be more intense in some cities, particularly those where interconnected global networks are helping blur the difference between the touristic and the everyday (Franklin 2003, p. 271).

The connectivity of global networks generates a two-way flow of people, images and information. Visitors and images travel from tourism generating regions to tourism destinations. Visitors return, bringing changed perspectives, behaviours and expectations. Products from the destination are sent back to the generating region or imitated there to add to consumption opportunities. On the one hand, a diversity of goods for consumption may add to the attractiveness of places but, on the other, globalisation and standardisation of consumption undermines local distinctiveness. Either way we become more sophisticated consumers of new places and residents have learned to 'act like tourists in their own cities' in their behaviours and consumption demands (Lloyd 2000, p. 7).

Globalisation has meant not only more flows of more tourists around the world, but that the rest of the world increasingly flows to us as products, images and new fellow residents. This makes for particular complexity in the process of touristification of life in the city. Residents' consumption demands are changed by their own experience as tourists; as Franklin (2003) argues, people learn new skills as tourists. Permanent and temporary migrants bring their own consumption habits with them, whilst tourists make their own demands on the city.

These changes in the consumption of goods and places, and the increased and intersecting mobilities that characterise contemporary urban living, challenge our ways of distinguishing between visitors and others. It becomes all the more important therefore to re-examine these interactions in considering how cities become desirable places to live, work and to visit. Understanding these interactions lies at the heart of this book. These changes in the tourist city are common to developed economies but we suggest they are possibly more intense in a relatively small group of world tourism cities. It is in these cities that changing relationships between visitors and others may be best observed and understood. Not only are these cities nodes on global networks, they are simultaneously major generators of tourism and other mobilities and leading destinations, receiving large volumes of visitors.

WORLD TOURISM CITIES

In this section, we need to expand our initial sketch of a world tourism city as multi-functional, attracting large flows of visitors, well located in global circuits with substantial historical assets, iconic buildings and that are also centres of cultural excellence. We draw on case studies from five such cities, but we could have drawn on others that share these qualities, cities such as Chicago, Toronto, Los Angeles and San Francisco in north America, and Madrid and Moscow in Europe. Distinguishing categories of city by size or function can lead to arbitrary selections and questions about comparable evidence. Such difficulties are evident in the substantial literature that has developed over the past twenty years that attempts to classify and define the roles that cities play on a global stage. Much of the debate about the 'global city', 'world city', 'global cultural city' or 'multicultural city' is contested and controversial. However, without wishing to enter such a contested terrain, we consider that the debates about world cities can help us clarify some of the features of the cities we examine in this book.

As we have discussed, at the centre of debates about contemporary globalisation are ideas about the connections between places and 'global flows'. There are many attempts to separate out different types of city from their positions in global networks. The 'global city' has particular economic functions (Sassen 1991), the 'gateway city' manages flows of people and the relationships between economic regions (Short, Breitbach *et al.* 2000), the 'global media city' captures specific economic sectors (Krätke 2003), and the idea of a 'global cultural city' claims to reflect success in the competition of images, branding, and mega-events (Yeoh 2005). The lists of cities and positions in the rankings are not static as economic fortunes change and, for example, new mega-projects upset old hierarchies. Numerous studies, reports and academic works have attempted to capture the dynamics of global competition between cities. What is important for our discussion is that lists of 'best city to do business', 'blockbuster exhibitions', or 'most creative city' point to the multiple registers on which cities have to compete. We suggest that cities that score well across the range of activities measured by world city watchers (see in particular the Globalisation and World Cities Group, lboro.ac.uk/gawc) have particular advantages. Multifunctional cities that accommodate international business, national élites, nationally and internationally important cultural assets, and can boast some distinctive historical monuments or iconic buildings, have substantial advantages in global flows of visitors, new residents and business investment. The cities we examine in this book all mix these functions in distinctive ways. Some have a more continuous history of economic and cultural dominance, and we see Berlin starting again as an important economic, cultural and political centre.

Studies of world cities often list economic assets and flows through airports as key indicators of importance in global networks. This may reflect the relative ease of access to such information. It is both frustrating to the researcher, and ironic for an industry that claims to be the biggest in the world, that hard information about tourist numbers is difficult to find and comparisons between cities almost impossible to quantify. Even at the national level, tourism data leaves a lot to be desired. Whilst the data available generally complies with UNWTO definitions, there can still be problems in making comparisons over time or between countries. Even basic data on visitor numbers and origins may be compiled on a sampling basis and be complicated by changing methodologies (for example, in the UK, over treatment of visitors from the Republic of Ireland). Problems at the local level are even worse, and attempts to use national level data to estimate tourism numbers in even large cities is hazardous. A recent review of UK tourism statistics by the Office for National Statistics revealed a series of weaknesses. For overseas visitors, it found that whilst 'national results from the International Passenger Survey . . . are of proper statistical quality . . . the regional results [which include London] are not' (Allnut 2004, p. 7). Problems with measurement of domestic tourism are greater still and the report found that the main data source, the UK Tourism Survey 'is not suitable for the production of tourism statistics . . . for areas smaller than the UK' (Allnut 2004, p. 7).

Individual cities do collect their own data on tourism but this is variable in quality, coverage and consistency. Whilst attempts are made to accord with UNWTO definitions, there are a series of difficulties in compiling even apparently basic data on a consistent basis. These include:

1   Visitors counted. Some cities (for example Paris) derive data on visitors from records of people staying in serviced accommodation. This provides a very comprehensive account for that sector, but ignores people staying elsewhere. Since world tourism cities are cosmopolitan, and the focus of many mobilities, they attract many tourists who are Visiting Friends and Relations (VFR) – and are likely to stay with their friends and relations rather than in serviced accommodation. Similar problems may occur with health tourists and educational tourists – for example those coming on a short language course and staying *en famille*. World tourism cities also attract many business visitors, but there is anecdotal evidence that many large companies now keep apartments for short term business visitors, rather than using hotels. For all these reasons, the number of visitors may be undercounted.

2   Areas to include. City boundaries are clear, but if tightly drawn some tourists whose main purpose is to see the city may stay outside the boundaries.

3   Day visitors and city residents. Day visitors do not stay overnight and so are not counted as tourists in UNWTO definitions. However, they may be drawn by the same city qualities as overseas and domestic tourists and visit the same attractions and use the same facilities (with the exception of serviced accommodation), and the numbers involved can be very large. Estimates for this group are of variable quality and many cities do not count them at all. Similar points could be made about 'internal tourism' – that is visits to other parts of the city by city residents.

4   Range of data. Given the difficulty in assembling information even on the apparently simple topic of visitor numbers, it is unsurprising that there is a dearth of data on more detailed matters such as purpose of visit, whether or not tourists have visited the city before, and which areas

they visit. This presents considerable difficulties in examining visitor behaviour.

All in all, these are formidable difficulties. They are compounded by the fact that city data is generally collected not by statistical agencies, but by local tourism organisations that have legitimate concerns to assemble data that is most useful in city promotion and marketing, rather than to present a statistically complete picture. The result is that comparable data for world tourism cities is confined to visitor numbers – and even so, there are inconsistencies about which visitors are included. The recent data on visitor numbers is shown in Table 1.1. More detailed information on visitation and on visitor perceptions is discussed in the individual city chapters, but cannot be accurately portrayed in comparable form.

Even these basic data shed some interesting light on tourism trends in world tourism cities in the twenty-first century. First, in general, tourism numbers, and particularly numbers of foreign tourists, are continuing to grow. With the crises that tourism has faced in recent years – terrorism, health scares, natural disasters – this is an important point. It is particularly notable that foreign tourists continue to be drawn to world tourism cities despite burdensome restrictions on air travellers, and the fact that they seem to present tempting targets for terrorists. Second, even basic data present multiple opportunities to claim ascendancy, which cities can and do take advantage of in promotion efforts. New York City attracts most visitors but London attracts most *foreign* visitors and Berlin has enjoyed the fastest *growth* in foreign visitor numbers. Third, local differences matter. London has seen a substantial fall in domestic visitor numbers whilst those from overseas have grown substantially. We can speculate that the lost domestic visitors have taken advantage of lower air fares to visit other cities in mainland Europe, including Paris and Berlin. Sydney also seems to be losing domestic visitors. Paris's performance may be understated by how boundaries are defined and, like Berlin, visitors are included only if they stay in serviced accommodation. These local dimensions are discussed in the chapters on each city.

At the least, we can say that tourism is a significant activity for our group of cities. One of the weaknesses of the current debate about the global city is the overemphasis on flows and in particular their economic worth. The place of Tokyo, London or New York in the global economy tends to dominate our understanding of world cities. However, some critics argue that we need a much better understanding of history and culture to make sense of the roles these cities play in the contemporary world (Abu-Lughod 1999). Too much emphasis on economic globalisation underplays the historical and cultural factors that make places. Some would argue that with its Roman origins, London has always been a world city. New York was always an international trading centre and in a very short time become the dominant east coast port and the landing point from which European culture disseminated into north America. Chicago has its roots in continental and international trade and

Table 1.1 Tourist visits to world tourism cities

1.1.1 Total visits

'000s	2000	2001	2002	2003	2004	2005	2006
Berlin*	5,000	4,930	4,750	4,950	5,930	6,470	7,070
London	31,650	28,350	27,700	26,000	26,200	24,600	26,160
New York City	36,200	35,200	35,300	37,830	40,000	42,600	43,797
Paris*	14,810	n.a.	14,830	13,970	15,190	15,400	n.a.
Sydney	n.a.	10,931	10,656	10,293	10,352	9,986	10,473

Notes:
*Figures refer to paid accommodation only

1.1.2 Domestic visits

'000s	2000	2001	2002	2003	2004	2005	2006
Berlin*	3,790	3,770	3,550	3,670	4,280	4,510	4,750
London	18,500	16,900	16,100	14,300	12,800	10,700	10,960
New York City	29,400	29,500	30,200	33,030	33,800	35,800	36,540
Paris*	5,620	n.a.	5,840	5,840	6,440	6,390	n.a.
Sydney	8,293	8,355	8,224	7,982	7,829	7,366	7,842

Notes:
*Figures refer to paid accommodation only

1.1.3 Overseas visits

'000s	2000	2001	2002	2003	2004	2005	2006
Berlin*	1,210	1,160	1,200	1,280	1,650	1,960	2,320
London	13,150	11,450	11,600	11,700	13,400	13,900	15,200
New York City	6,800	5,700	5,100	4,800	6,200	6,800	7,257
Paris*	9,190	n.a.	8,990	8,130	8,750	9,010	n.a.
Sydney	2,423**	2,576	2,432	2,311	2,523	2,620	2,631

Notes:
*Figures refer to paid accommodation only
**Refers to period July 1999/June 2000

Sources:
London Visitor Statistics (2007), VisitLondon.
NYCVisit <http://www.nycvisit.com/content/index.cfm?pagePkey=57>
OTCP (2004), Chiffres-cles du Tourisme à Paris 2004, Office du Tourisme et des Congres de Paris
OTCP (2005), Chiffres-cles du Tourisme à Paris 2005, Office du Tourisme et des Congres de Paris
OTCP (2006), Chiffres-cles du Tourisme à Paris 2006, Office du Tourisme et des Congres de Paris
OTCP (2006b), Point Presse du 31 Août 2006, Office du Tourisme et des Congres de Paris-Observatoire du Tourism Parisien
Statistisches Landesamt Berlin
Sydney Region Tourism Profile Year ending December 2000, 2001, 2002, 2003, 2004, 2005, 2006. Tourism New South Wales
Tourism Australia (2007), Visitors Arrivals Data, http://www.tourism.australia.com accessed on 17 July 2007

invented and reinvented its image through 'world's fairs'. Those legacies are still vital parts of these 'world' cities. Other, once dominant, centres of world trade, Rome, Istanbul and Cairo have lost this economic role but retain the heritage and cultural assets that draw large numbers of visitors.

We see an important economic role both in the past and the present as an essential feature of the cities we examine. In north America, we might include San Francisco, Los Angeles, Chicago and Toronto in our category of world tourism city and in Europe, perhaps Madrid and Moscow. Smaller cities, like Barcelona and Prague, are tourism hotspots but lack the multi-functional, polycentric scale of other cities. Neither do Rome or Athens have the economic importance of London and Paris and the transitional Berlin. As a world tourism city, Chicago, for example, is large, polycentric and multifunctional with substantial and varied visitor attractions. Through transformation of its downtown and lakeside areas, Chicago demonstrates (as an exemplary model according to some, see Clark 2003) the process through which the tourist city and 'entertainment machine' come to play a vital econ-omic role. A study of neighbourhood transformation in Chicago (see Lloyd 2005) shows some of the interactions between tourists and residents that we examine in later chapters. It is this accumulation of roles and activities that distinguish world tourism cities.

For geographers, classifying types of world and global city has been chal-lenging and contested. However, whatever caveats academics may put on their work, city leaders have been very keen to take up the language of the world city and deploy and develop world city assets in competition with others. New York sees no problem in advertising itself as the 'tourism capital of the world'. We need to appreciate the very important role that ideas of world and global cities play in public policy. Competition is most visible in the tall buildings that signal global city ambition. However, competition also registers in other sectors of public policy. The right mix of cultural assets can also define world city aspirations. Cultural policy also favours notice-able buildings, and many cities with economic claims to global status have also recently invested in prestigious and expensive architecture. In east Asia, such 'cultural imagineering' has sometimes been contested (Yeoh 2005 and Kong 2007) and tended to create separate cultural quarters to manage visitors and which offer limited representation of local cultures. Other cities seek to market their global status not through prestige projects, but through the qual-ities of their cultural outputs. Thus, for the Mayor, London is a 'world-class city of culture' and:

> The wealth of London's creative industries could be repackaged as a specific attraction including design, fashion, digital media etc. The growth of guided and self guided walks could be encouraged exploring new parts of the city and fresh themes such as distinctive local sounds, oral history or developing new interactions using mobile phone technologies.
> (Mayor of London 2004a, pp. 68–69)

Whatever the actual strength of these economic sectors, city leaders believe that culture and creativity can be repackaged to commodify new parts of the city and enhance city image in relation to global competition. According to its Mayor, 30 percent of business visitors cite art and culture as reasons for staying on in London. London claims unparalleled assets and needs to 'Develop its brand and promote itself as a world cultural city and tourism destination' (Mayor of London 2004a, p. 32).

City marketing has recently taken another look at cultural representation and world cities have increasingly become defined by multiculturalism. Cultural diversity is assumed to be a 'sufficient and necessary' condition of cosmopolitan character and a vital part of the claim to world city status (King 2006, p. 322). London's cultural policy picks up this theme, claiming that with 300 languages spoken in London schools, the city is 'one of the most multicultural cities in the world' (Mayor of London 2004a). However, as King 2006 argues, no city can ever hope fully to represent the global diversity of languages and cultures. What is important, is the way diversity is valued in branding the city in competition with others. Diversity is often represented as a city-wide cultural quality but without appreciation of the spatiality of diversity within the city or the relative strengths of local communities. In London, cultural policy would like to exploit cultural diversity in Brick Lane, Brixton and Southall but without appreciation of the many problems of discrimination within the cultural economy (Talbot 2004).

The marketing of the 'multicultural city' has also to keep pace with rapid urban change. The social and spatial structure of New York City has always reflected its history as a home to migrants. However, Minnite (2005) argues that recent international migration has become globalised and the city's former distinctive multicultural character has become lost in 'hyperdiversity'. The city's global connectivity has changed and so have the characteristics of neighbourhoods off the beaten track. Middle-class European visitors who may have enjoyed 'slumming' in Manhattan a hundred years ago would find the slums have all but disappeared through the relentless gentrification of the city. 'Hyper' migration and 'hyper' diversity create new relationships between the economic, social and spatial core and the periphery. Visitors have to learn to negotiate the changing urban structure. World tourism cities are changing, and economic, social and cultural changes impact on the draw of traditional attractions and the potential and desire to move off the beaten track.

Recently, globally important cities have become more visibly concerned with security. Savitch and Ardashev (2001) asked if in the contemporary world cities had become 'targets' for terrorism. Questions about safety and security clearly impact on the choices of visitors. For example, in post 9/11 New York, new developments in 'patriotic' tourism and re-branding of the city (Greenberg 2006), have to be located in a broader consideration of urban security. Graham (2004) and others argue for a broad appreciation of relationships between city and security and surveillance. If big cities present bigger targets, then we need to ask how far visitors are prepared to move outside rings of surveillance. The central area of London is included in 'one of the most daunting defence systems protecting a major world city' (Coaffee 2003, p. 194). In 2006, the video surveillance of the congestion charge zone was extended to cover the museums quarter. On the other hand, critics of the sanitisation of cultural quarters (see Reichl (1999) on Times Square, for example) may see increased security as an incentive to move off the beaten track.

Images of cities change, as do the imaginary boundaries between traditional landscapes and newly discovered parts of the city. The importance of history in defining world tourism cities is visible in accumulated landmarks, cultural assets, 'zones of prestige' (Maguire 2005) and the routine of celebratory events. For many visitors, the cities examined in this book are well understood and their iconic tourist landscapes easily negotiated. In the first half of the twentieth century, New York's 'vertical landscape' gave the city the instant recognition already enjoyed by many older European cities and spectacular buildings elsewhere. Sydney's iconic landscape is also a twentieth century product. Berlin's landscape underwent distinctive periods of transformation and, importantly, interpretation and reinterpretation (Neill (2004) and Lehrer (2006)). The strong, well known images of these cities represent complex legacies, and the relationships between familiar and less familiar parts of the city are continually renegotiated.

An important part of our understanding of world tourism cities is therefore how visitors interpret different parts of the city and how impressions change over time. City image is not just a set of impressions of iconic buildings. Visitors have always been impressed by the liveliness and variety of the big city. Porter (2000, p. 35) describes the attractions of Georgian London as being the 'activity, news, personalities, fashion, talk, discussion, and nonstop parade'. Two centuries later the 'crush and rush of the city had established itself as one key marker of the city's world importance' (Gilbert and Hancock 2006, p. 81). The intangible qualities of modernity mark out important world cities. More tangible modernisations have also added to the historic attractiveness of cities. Visitors to Paris in the nineteenth century came to view the new sewers and abattoirs. In New York, modernity brought other attractions. By the early 1900s the city had an impressive new urban infrastructure, such that: 'These sights suggested a different kind of tourist city, one in which the spectacle of mass movement, record-breaking engineering, and even the infrastructure of the modern city became primary attractions' (Gilbert and Hancock 2006, p. 85).

At different times visitors interpret and reinterpret the attractiveness of cities and the places within them. In our group of cities, as a result of their multifunctional character and polycentric structures, visitors (and others) seem to enjoy both tradition and modernity. The 'spectacle' and 'crush and rush' take visitors both to and away from traditional and familiar landscapes. Some visitors make distinctions between 'travellers' and 'tourists' and create itineraries beyond traditional landmarks in order to confirm their prejudices.

This may push some visitors off the beaten track, as in the past, to the Paris abattoir or the slums of twentieth century Manhattan. Such contrasts, and such opportunities, are more available in large polycentric, multifunctional cities whose histories of tradition and modernisation have in the past created the potential for distinctions between regular and irregular itineraries.

Gilbert and Hancock (2006) argue that the transformation of the city into an object of global consumption creates the opportunity for reaction to standardised and commodified paths through the city. We suggest these processes are more marked in our group of cities, cities that have been through the transformations of modernisation and where their history and urban form create both a well known and relatively stable commodified landscape of tourism *and* other opportunities to negotiate new paths. We can see some images of the city as being relatively fixed but we have to acknowledge that this is the outcome of the work of visitors and others as they attempt to stabilise global and local linkages and the diversity of spaces in the city.

Appreciation of the structure of the (polycentric) city depends on the point of view. The 'global' and 'local' terms that define the many categories of global financial, media or cultural city cannot be taken as representing fixed places. Moore (2004) argues that the 'social maps' through which people organise their lives deploy different and changing ideas of the global. Culture and consumption mark out differences between cities. For example, German bankers in London saw Frankfurt as offering less high culture and expensive shopping than London, and the shopping and entertainment opportunities of London offered a means through which they could define themselves as 'cosmopolitan world travellers' (Moore 2004, p. 8). Visitors, as well as businessmen, make and remake such connections and interpret the global position of the city.

The world tourism city through its history, its multi-centred and multifunctional structures, presents opportunities for interpretation and reinterpretation, and we would expect different actors to take different views and for the relatively stable, commodified landscape of tourism to be reconfigured by visitors, residents and others as they negotiate these big cities. We define world tourism cities as multifunctional and polycentric with the capacity to draw visitors off the beaten track and where visitors and other city users may share in the creation of new tourism places.

WHAT DO VISITORS TO THE WORLD TOURISM CITY WANT?

At the heart of our idea of the world tourism city is the interaction of visitors and other city users and the active role of visitors in contributing to new urban experiences. We can understand more about visitors and their motivations by turning to some of the insights of work on tourism more generally. Tourism studies have generally depended on a model in which the visit arises from a combination of 'pull' and 'push' factors. Pull factors constitute the attractions of a particular destination – the reasons that a potential tourist might want to visit. These include the material and tangible (for example the Eiffel Tower, Statue of Liberty, or Sydney Harbour Bridge) but also the experiential (life in the city) and the possibility of increasing the visitor's cultural capital (for example by visiting a fashionable destination). Push factors are those that make individuals want to travel. They have been examined in psychological, economic, anthropological and social terms (Sheller and Urry 2004) and have led to the construction of a variety of tourist typologies. These seek to categorise visitors on the basis of their central motivations or consumption demands. Cohen's (Cohen 1972 and 2004) pioneering typology distinguishing a range from the drifter to the organised mass tourist, is one of the earliest attempts and remains well known. Subsequently, other perspectives have been developed including those focused on the gaze (Urry 1990), changing preferences of the 'new tourist' (Poon 1993), the self aware post tourist (Feifer 1986) and varieties of cultural tourist (McKercher 2002) – see Harvey and Lorenzen (2006) for a recent review. Newer perspectives have emphasised the need to avoid relying on single explanations for complex behaviours, and that tourists have multiple goals, some of which may appear contradictory to observers. Williams and Shaw (1998) point out that tourists may seek an authentic experience but also revel in consuming what they realise to be a staged and non-authentic experience, whilst Cohen (1995) sees postmodernism as sanctioning enjoyment of contrived attractions. Recent work in tourism has emphasised interconnections between pull and push factors as visitors contribute to changing the cities they visit (for example Franklin (2003); Maitland and Newman (2004) and Peel and Steen (2007)).

These developments suggest we need to avoid thinking of visitors as 'dumb tourists' (McCabe 2005) or 'passive consumers or even cultural dupes . . . easily fooled' (Meethan 2001, p. 112). Tourists can be discriminating and subtle in the way they experience places, but this is frequently not reflected in tourist typologies – perhaps because 'attempts [that] have been made to subdivide the tourist by various typologies . . . neglect the tourists' voice' (Wickens 2002, p. 834). Equally, we must not see visitors as synonymous with leisure tourists. Leisure visitors are only one element of tourists in the world tourism city and we need to pay attention to the range of other tourists coming to the cities, many of whom have some connection to the city before they make their visit – either because they are visiting friends or relations or travelling on business to meet colleagues. More broadly, there are 'many overlaps between "tourism" and other kinds of business, professional and migratory movement' (Sheller and Urry 2004, p. 6). World tourism cities attract many different types of tourist but their roles as centres of business, governance, research, education and culture mean they also attract temporary migrants. For example, professionals in transnational corporations and financial institutions may have short-term contracts or temporary assignments; students come to study for degrees and academics for exchange or research projects; 'creatives' arrive to make movies or devise campaigns.

Fainstein, Hoffman *et al.* show how this mix of mobilities arises from global changes that have generated:

> a greatly enlarged, educated group of consumers, drawn primarily from wealthy countries but also from the middle and upper strata of developing nations, who travel for business and pleasure; migrants and immigrants who fill low-wage jobs in the expanding urban service sector and make frequent trips back to their home countries; students and drifter tourists who manage on low budgets.
>
> (Fainstein, Hoffman *et al.* 2003, p. 243)

Changes in the global economy have simultaneously seen dispersal of production and consumption, requiring more business travel for sales and marketing along with the expansion of industries and occupations requiring high levels of education. Some argue that this has led to the growth of a 'cosmopolitan consuming class' of frequent travellers and temporary migrants with 'sophisticated tastes [and a] . . . demand for cultural amenities' (Fainstein, Hoffman *et al.* 2003, p. 243). World tourism cities are particularly likely to attract members of the cosmopolitan consuming class, both because of their concentrations of global economic functions and their concentration of consumption opportunities for the well off and sophisticated.

In short, life in the world tourism city is lived by residents, who increasingly behave like tourists in their own city, by short term migrants, who will presumably be interested in exploring the city whilst they are there, and by a variety of different tourists, many with strong connections to the city. This is a complex mix of 'intersecting mobilities', and it is often difficult to see how we can make clear distinctions between 'tourists' and other types of 'city user' (Martinotti 1999) in the use they make of and the effect they have on the city.

However, despite these compelling arguments for a focus on motivation rather than fixed typologies, there has been comparatively little research effort devoted to discovering what visitors want. The visitor experience is at the heart of travel and tourism, and depends on their perceptions and behaviour: as Ashworth and Dietvorst (1995) remind us how tourists as consumers manage their own holiday experience is of critical importance (Ashworth and Dietvorst 1995). Yet we still know little about what draws people to cities or about their experience once they are there (Page 2002). Most research has examined tourism development from a supply-side perspective, concerning itself with place production rather than how and why places are consumed (Selby 2004). In effect, this means that discussion on tourism in cities tends to be rooted not in the visitor's perspective but the view of those who produce places – developers, planners and tourism marketers for example. This reinforces the stereotype of the dumb tourist, a passive and reactive consumer of experiences designed by others. As Wickens (2002, p. 834) says, 'much writing treats the tourist as a unitary type confined to a tourist bubble'. This oversimplification undermines attempts to unravel the complexity of tourism in cities. However, despite recent work to investigate the characteristics and experience of visitors to particular areas in cities (Maitland and Newman 2004, Selby 2004, Hayllar and Griffin 2005 and Maitland 2007b), there is still 'limited research material . . . focusing on the experience of tourists within [tourism precincts and] . . . an understanding from the tourist's perspective has been a neglected dimension' (Hayllar and Griffin 2005, p. 518).

Thinking about the tourist experience of world tourism cities, as visitors move off the beaten track, links to ideas about authenticity, the accumulation of personal cultural capital, and the role that others have in the construction of an individual's experience – whether other tourists or others from the range of city users. A substantial debate about the motivation of tourists has developed around the tourist's assumed search for authenticity since MacCannell's (MacCannell 1976) pioneering work (see Wang (1999) for a review). Gilbert and Hancock (2006) noted how those European visitors to New York in the early twentieth century, who sought experiences off the beaten track, were also quick to complain about the lack of authenticity in their experiences as New Yorkers 'staged' events. Concerns about authentic urban experiences have long been pursued in the planning, urban design and architecture literature in debates about the conservation and remaking of cities (for example Lynch 1960, Jacobs 1961, Venturi, Brown *et al.* 1972 and Allmendinger 2001). The search for the authentic is linked to an interest in the pre-modern and in 'heritage', often sought in the developing world; but we can also identify a concern to distinguish the authentic and 'real' from the inauthentic and fake, or in a desire to get behind the scenes and have a 'real' experience.

These discussions can become confused through the plurality of meanings ascribed to the notion of authenticity (Wang 1999). Whilst authenticity can be objective (as when a museum authenticates the provenance of items in its collection) it can also be socially constructed or derive existentially from the visitor's experience. For Hall (2007) what counts is connectedness and experience:

> the notion of authenticity should not be used with respect to things or places. Authenticity is instead derived from the property of connectedness of the individual to the perceived everyday world and environment . . . Authenticity is born from everyday experiences and connections which are often serendipitous.
>
> (Hall 2007, p. 1140)

This links to debates about consumption as a search for experiences, and the development of the 'experience economy' (Pine II *et al.* 1999). Pine (2004) says that 'as the experience economy matures a shift is identified toward authenticity' (cited in Yeoman, Brass *et al.* 2007, p. 1128). Indeed, Yeoman, Brass *et al.* argue that a series of trends is shaping the 'authentic tourist' (Yeoman, Brass *et al.* 2007, p. 1130) who will be affluent and well educated, individualistic and experienced in multi-cultural contexts – in other words, apparently with the characteristics of the 'cosmopolitan consuming class'.

The 'authentic tourist' is a contested notion, and May (1996) shows that mass tourists as well as members of the cosmopolitan consuming class make complex judgements about what is a 'real' or authentic experience. Whilst much of the time they may prefer the experience of the resort or tourism zone, at times they may want to 'attempt to move beyond the usual tourist experience towards a back-stage encounter' by seeking out or stumbling upon an 'unusual little place' even in a well established resort (May 1996, p. 729). Members of the cosmopolitan consuming class, conscious of the increasingly commodified landscape of tourism, give higher priority to finding the real and authentic, and to experiences that are not engineered for tourists. However, for both groups, experiences that are distinct and unusual are important in the accumulation of personal cultural capital. In world tourism cities, getting off the beaten track and feeling connected to the everyday can provide such 'real' experiences.

Experiences derive not simply from place attributes, but also from other place users. Harvey and Lorenzen (2006) argue that whilst it is acknowledged that visitors play a variety of roles, too little attention is paid to the importance of other tourists, who are needed if the roles are to be played out. They term this the requirement for co-tourists. Co-tourism can occur when part of the tourist experience is meeting and interacting with other visitors with similar cultural capital. They cite the example of wine lovers visiting the Santa Barbara area of California, whose tourism experience long depended in part on the pleasure of meeting fellow connoisseurs. Since the release of the recent film 'Sideways', there has been a new influx of tourists who wish to see the locations of the film, eat in the featured restaurants, and act out scenes: meeting other visitors who are doing the same adds to their very different experience. In both cases the presence of co-tourists is an important element.

If we apply this notion to the world tourism city, we can see that it is one aspect of the way in which other city users can be central to the tourist experience. The presence of other visitors with similar tastes can add to the attraction of areas – a point long recognised in the case of backpacker enclaves in cities, for example (Peel and Steen 2007). However, the broader point is the way in which the presence of other city users with similar cultural capital can be central to the tourist experience whether they are 'local people', and or other visitors. Members of a 'cosmopolitan consuming class' may value interaction with one another as much as wine lovers, and seek out areas of the city where they can engage in shared consumption with 'co-city users', whether visitors or residents.

VISITORS, RESIDENTS AND SHARED URBAN PREFERENCES

In this section, we explore further the idea of shared interests and preferences between visitors, residents and 'co-city users'. Middle-class preferences have dominated the recent development of urban tourism. These preferences are summed up by Glaeser *et al.* (2000). First, the 'rich variety of services and consumer goods', second, 'aesthetics and physical setting', third, 'good public services', and fourth, the ability to move around quickly (Glaeser *et al.* 2000, p. 28). Some 'good public services' (social services for example) may be of little interest to visitors but the range of other services and quality of life may be more generally valued.

Aesthetics and physical setting clearly have the power to draw resident and visitor alike to some parts of the city. Over the past forty or more years, studies of the process of residential gentrification have analysed the appeal of particular landscapes. Of particular interest in gentrification studies is the emphasis on the role of new residents in re-evaluating the aesthetics of often disregarded city landscapes, whether Georgian terraces in London, or industrial lofts in New York. The urban middle-class assigns values to styles of architecture and to the social character of neighbourhoods. Gentrification is understood not just as economic process, but as involving a cultural dimension as gentrifiers select and rebuild city quarters (Jager 1986). As residents build cultural capital, so we can see tourists working in a similar way in marking out destinations and places to be and possibly sharing the work with other city users.

More recently, gentrification research has moved beyond the particular local circumstances of social and economic change in few cities and begun to emphasise the connections between cities through the idea of a, possibly global, gentrifying class. Work in this field of urban studies clearly overlaps with the debate about urban tourism we reviewed above. Lees (2007) for example, notices the 'the interconnection of gentrified neighbourhoods transatlantically' through the behaviour of British migrant workers in Brooklyn (Lees 2007, p. 228). Clearly, some cities, because of their economic and other roles, have the ability to attract and retain an international workforce and this internationalised labour market has visible impacts on particular neighbourhoods. Strong ties between cosmopolitan élites influence processes of gentrification (Atkinson and Bridge 2005). We can imagine a cosmopolitan class with 'skills that now transfer anywhere' with portable social resources and strong cross-national ties (Atkinson and Bridge 2005, p. 9).

If, in some well connected large cities, gentrifying areas respond to the expressed demands of a self-defined, translocal class, then we should expect overlap with the desires of some visitors. In Sydney, for example, Rofe (2003) identifies a transnational élite making consumption demands and expressing a desire to be seen to be 'global'. A cosmopolitan orientation demands restaurants and galleries where distinctions of taste can be displayed. Gentrified areas of London '. . . have become global spaces, serving the international service class diaspora in a safe environment that acknowledges the cultural capital of the consumer' (Butler 2007, p. 184). In London, New York and Sydney, similar studies are developing this idea of global, gentrifying class and some researchers argue that housing market investment is informed by relativities between New York, Sydney and London (Atkinson and Bridge 2005, p. 10). In the cities we examine in this book, gentrification has a particular importance.

The arguments about a 'global gentrifying class' have some value but there are some important questions. Is this global process the same in all cities? Should we see gentrification as being driven by a metropolitan or cosmopolitan mentality or by more locally specific economic and social contexts and public policies? A more sceptical approach (Bridge 2007) to the idea of a global gentrifying class argues that different aesthetics and local processes are still important in remaking city neighbourhoods. Whatever the claims about a global class in these cities, the gentrification process has similar impacts, valuing some neighbourhoods over others and managing inner city life through the market (Butler 2007, p. 183) and demanding a range of high quality consumption opportunities. Gentrifiers demand what Glaeser *et al.* (2000) label a 'rich variety of services and consumer goods' that can be equally attractive to other city users.

This shared interest in consumption may be shared by a particular gentrifying class, transnational workers and visitors who have aesthetic values and consumption demands in common and, where they can, always make places feel like home. However, we should remember that as gentrifiers mark out neighbourhoods, other city residents may be displaced and particular cultural values imposed on multicultural cities. Atkinson and Bridge (2005) refer to a new 'urban colonialism' as a dominant group imposes itself among the multiple demands and cultures of the contemporary city. Gentrification research has recorded stories of conflict over neighbourhoods and the path of a middle-class expansion through the inner areas of large, multicultural cities may not be smooth. These lessons from work on gentrification may inform our understanding of how the uneven urban landscape is negotiated. Middle-class urban preferences may be limited and as Butler observes in north London: 'There is no sense here of the middle classes being embedded in a more "authentic", volatile or rounded London' (Butler 2007, p. 183). The 'post colonial', 'hyperdiverse' city gives a fluid context for an urban middle-class seeking to impose its taste on some parts of the city and for the negotiation of spaces on and off the beaten track.

Gentrifiers invest in, and revalorise, older residential landscapes but often also reappraise undervalued industrial and commercial areas. Fashions in desirable places change, from Victorian terraces, to old tenements, to mid-rise industrial buildings. These different landscapes and different residential preferences throw up possibilities for other activities, for example, inserting shops and cafés into the ground floors of converted tenement buildings or a mix of uses into converted factory buildings. The mix of residential with other uses has been an important dimension of many accounts of gentrification. There has been a long term concern to understand the pioneering role of arts and artists in defining new residential areas and with understanding how gentrification processes fit into a city of mixed land uses (Coupland 1997). Culture and shopping are important parts of the gentrifiers' market driven approach to urban living, and the character and quality of local consumption offer further opportunities to make and display distinctions of taste. Residential change and changing consumption evolve together. For example, over recent decades lower Manhattan has been revitalised and so local streets have become 'part of a recharged, transnational shopping economy' (Zukin 2005, p. 27). In some cases, where shopping has resisted suburbanising and standardising trends, neighbourhood shopping has become a city-wide attraction and a draw for visitors. Over time, neighbourhood shopping changes and becomes more or less attractive to potential residents or international shoppers. For example, over four decades, Covent Garden lost its local food and service shops to be replaced by new specialist shops that in turn were swamped by international brands. Residential change may lead to the loss, or arrival, of new attractive shopping streets. Relationships between commercial activities and residential markets are complex and may differ markedly from place to place. For example, in several neighbourhoods in Toronto, commercial activity, in particular the 'ethnic packaging' of cultural products, was used to activate processes of gentrification (Hackworth and Rekers 2005).

In some cities visitors are contained, but in others they can 'escape the confines of enclosure' (Judd 2003, p. 30). That they should want to escape has to do with several overlapping forces. The world tourism city offers multiple experiences of consumption and entertainment, and new urban spaces that can be imagined as objects of desire (Blum 2005). In some cities, tourism can be seen as a 'natural by-product' (Judd 2003) of a new planning priority given to urban amenities and the neighbourhood scale. In this way, we could see the shift of emphasis in Paris in the mid 1990s from big projects and urban renewal to more concern with residents' demands and emphasis on the local scale, on benches, green spaces, small fountains and traffic regulation, as creating new places that appeal to visitors as well. Some authors associate such policy shifts with long term shifts in social values that give rise to demands for public goods and concern for the amenity value of cities (Clark 2003). Clark's model is Chicago. We suggest similar forces are at work in a group of cities like Chicago, multifunctional and polycentric cities with historical assets and which offer co-city users opportunities to create and consume new places. These processes will not be without conflict. The preferences of middle-class co-city users, the demand for consumption opportunities and the repackaging of neighbourhoods may exclude others. There are consequences for public policy at city-wide scale and at the scale of local neighbourhoods as political leaders and communities may seek to extract benefits from urban tourism off the beaten track.

OUTLINE OF CHAPTERS AND THEIR THEMES

The chapters in this book take up these themes about tourism, gentrification and neighbourhood development and draw on new research into visitors' differing perceptions of areas within our group of world tourism cities. The chapters also examine the responses of public policy whether through city marketing and city development strategies or local attempts to re-brand cities and neighbourhoods. The core themes of the book are set in the context of a review of tourism trends and policy directions for each world tourism city.

We start in New York City (NYC) and Jill Simone Gross examines recent trends and issues in the NYC tourism market with a reconsideration of tourism as a driver of economic revival. The chapter considers the new tourism interest in Harlem, Queens and Brooklyn and, drawing on original research, examines the role of local agencies and local communities in the transformation of downtown Brooklyn. These attempts to re-brand neighbourhoods have to be seen in the context of a dominant Manhattan and well established images of the city and its attractions. The chapter explores the possibility of a 'dual' tourism market as some visitors are drawn to the re-branded neighbourhoods outside Manhattan.

The re-branding of neighbourhoods and larger urban quarters is also a theme of Patrizia Ingallina and Jungyoon Park's chapter on Paris. They look at recent developments in eastern Paris and in the suburbs where tourism has only recently attracted the attention of urban policy makers. Their discussion of new, off centre neighbourhood tourism is set in a wider context of debate in Paris about the competitive position of the city whose pre-eminence as a tourist destination has been challenged on a number of fronts. The chapter focuses on city image and rebranding alongside the growing awareness in public policy of the need to integrate economic development and quality of life objectives in new urban policies.

Like New York, London has been through some reassessment of its competitive position and responses to recent crises, including 7/7. Robert Maitland and Peter Newman's chapter examines approaches that have been taken to developing new attractions, strategies for branding and marketing and the integration of tourism with other aspects of city development. Original research with overseas tourists in Bankside and Islington – two areas where tourism development is new and growing – throws some light on the perception and re-evaluation of areas off the beaten track and suggests some lessons that can be learned for other tourism development outside central London in areas such as the East End.

In Berlin, Johannes Novy and Sandra Huning examine the growth of tourism since the city's re-establishment as Germany's capital. The chapter considers trends in off the beaten track neighbourhoods with a particular focus on Kreuzberg, which developed as a bohemian and 'alternative' neighbourhood when Berlin was a divided city and has become more popular with tourists following reunification. The chapter highlights questions about harnessing these trends and the possibilities of achieving sustainable, neigh-bourhood based, tourism.

In the next chapter, Bruce Hayllar and Tony Griffin use interviews with tourists in Sydney to understand the way that different visitors may perceive and use areas differently and how that relates to their wider tourist experi-ence of the city. They contrast off the beaten track areas with purpose designed tourism zones and draw conclusions about the reconciliation of differing demands and perceptions of different visitors in the same area and the implica-tions for tourism development and marketing.

We expect some similarities in the ways in which this group of cities responds to changing demand and continues to offer new, desirable destinations for experienced urban tourists. Our final chapter reviews similarities between the cities and we discuss how far the themes in this introduction resonate across old and new world cities and across continents. The governance of tourism and of urban policy more generally sits in very different traditions and we go on to discuss how in the different cities public policy may be adjust-ing to the ways in which visitors are interacting with locals off the beaten track. We began this chapter with a discussion of the difficulty of finding reliable information on urban tourists. Each of the chapters reviews tourism information for its own city and in the final chapter we return to the discus-sion of comparability. Each chapter takes a similar path through a discussion of general trends in the city and each then draws on original research to offer new insights into how this group of World Tourism Cities are offering new experiences to visitors.

2 New York tourism

Dual markets, duel agendas

Jill Simone Gross

INTRODUCTION

> New York is the concentrate of art and commerce and sport and religion
> and entertainment and finance, bringing to a single compact arena the
> gladiator, the evangelist, the promoter, the actor, the trader and the mer-
> chant. It carries on its lapel the inexpugnable odor of the long past, no
> matter where you sit in New York you feel the vibrations of great times and
> tall deeds.
>
> (White 1949, p. 19)

White was describing the New York of 1948, a city that offered both visitor
and resident alike a combination of elements that most perceived to be indicative of the 'authentic' urban experience. In New York, one could find a
real 'slice of life', in which diverse populations could rub elbows with one
another. It was this incredible variation that also made New York a magnet
for populations, business and visitors.

While in some regards, it can be argued that White's New York of 1948
continues to exist today, in the minds of others, the pressures of global markets have led urban leaders – political and economic – to promote the development of what Judd refers to as 'tourism bubbles'. These are areas that are
developed for business and tourist markets, often they are both physically
and functionally isolated, and in turn the visitor is separated from the
living city. In New York, a range of these areas have emerged over the past
three decades, for example, the South Street Seaport, a homage to New York's
maritime past, in the form of a semi-enclosed water front shopping mall and,
the World Financial Center, that mixes offices, waterfront, and shopping along
with a large glass enclosed 'ground zero' viewing area. The tourist visiting
these locations is captured within a defined activity zone – a place where they
can shop, dine and be entertained – without direct contact with the 'real city'
or, as Fainstein and Gladstone remind us, 'at their most extreme [these kinds
of], tourist destinations become wholly dis-attached from their social context' (Fainstein and Gladstone 1999, p. 27).

Alongside the purposefully constructed tourism location is an equally problematic homogenisation of goods and services, not simply affecting the visitor's experience, but also the resident. Indeed, some suggest that in combination, these tendencies threaten to 'squeeze every last drop of individuality from our communitie's (Silbert 2005). As one New York blogger comments:

> New York City, since the early '90s, is slowly being turned into a shopping mall. Soho, once well known for its galleries and interesting art spaces, is nothing but Bloomingdales (!) and Old Navy. At least three new megaplexes have opened, and there's of course the Disneyfication of 42nd Street.
>
> (Zandt 2005)

While there is no question that in certain parts of New York City, particularly Manhattan south of 96th Street, a sameness is slowly emerging, shaped by the policy drive to maintain global position, there is also a more locally driven set of policy agendas. These are driven by efforts to preserve and promote distinct neighbourhood identities in communities outside the centre in Upper Manhattan, Brooklyn, Queens, the Bronx, and Staten Island. The result is the emergence of a 'dual' tourism market and 'dueling' policy agendas of globally versus locally driven political interests. This duality and 'duelity' is supported not only by supply side forces, but also by an emerging set of demand side pressures, derived from visitors seeking to experience the 'real' New York.

This chapter presents data on the changing contours of New York City tourism and the tourists it attracts, both prior to, and in the aftermath of 9/11.[1] It will explore two visions of the 'real' New York, the global and the local, and its impact on the way tourists experience these different parts of the world city. In what follows, I will first provide an overview of the changing contours of tourism as an economic development strategy. This is followed by a discussion of New York's 'dual tourism' market and its 'duel tourism' politics, focusing on the underlying tensions between centre and periphery and efforts to re-reveal the 'authentic' face of New York's neighbourhoods. This leads us to a discussion of three 'new tourism areas' on the periphery of the Mid Town Manhattan, located as Figure 2.1 below shows, in Downtown Brooklyn, Long Island City in Queens and Harlem in Upper Manhattan, and the tourists they attract. It is in these areas that a growing number of New York's visitors can now be found:

> Not the must-see sights, but the mundane ones . . . The unexpected truths in the most ordinary experiences . . . salsa concerts in the Bronx's Orchard Beach . . . bus rides through Prospect Park in Brooklyn . . . wander[ing] Harlem's historic streets . . . [To] . . . catch the real vibe of the city.
>
> (Swarns 1996)

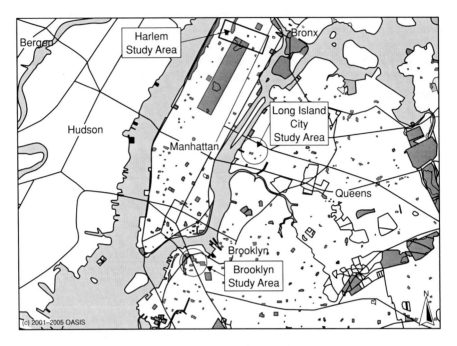

Figure 2.1 New York City showing new tourism study areas
Source: NYC Oasis, 2008

NEW YORK TOURISM – AN EVOLVING ECONOMIC DEVELOPMENT STRATEGY

In general, the resources required for the promotion of an area as a tourist destination are diverse: architecture, history, shopping, culture and political advocates are but a few. The tourist market is equally complex. On the supply side are direct (those offering goods directly and specifically for the tourist) and indirect (those providing services for both resident and tourist alike) service providers. The demand side includes those coming to the City for business, education and leisure, from both domestic and international locations. World city tourist destinations, like New York, therefore benefit from being multi-functional centres for work and play (hubs of finance, travel, culture and communication).

Up until the late 1970s, tourism in cities like New York tended to be viewed as a byproduct of natural geographic advantage, local economic and infrastructure investment, national policy and global economic shifts. In turn, local policy makers often viewed tourism as the indirect result of other forces and policies, though, not as its own policy area, *per se*. New York City's tourism policies tended to be expressed indirectly through taxation and finance programmes supporting economic development and entrepreneurship.

When direct investment came, it was targeted towards promotions for domestic markets. In 1977, for example, the 'I Love New York' campaign was launched with \$4.3 million initially to build awareness of New York State as an outdoors destination for recreation (Feron 1979). The marketing campaign is credited not only with reversing a decade long decline in tourism statewide, but also in producing an increase in domestic tourist spending throughout the state of almost \$300 million. In 1979, the possibilities of utilising tourism as an economic development strategy were becoming increasingly evident, as was the need.

Deindustrialisation, suburbanisation, recession and the oil shocks had left many American cities in distress by the mid 1970s and New York was no different. Between 1950 and 1970, the central city experienced both population and manufacturing job losses (Gross 2005). Alongside of this, the City also experienced rising crime rates. By the mid 1970s, the City was on the verge of bankruptcy and a lack of resources also led to declining services – dirty streets, graffiti covered subways, and a general feeling that the City was unsafe.

Not surprisingly, tourism suffered – the City was perceived then as a dangerous place. Economic restructuring was further complicated, by declining fiscal support from federal government. The result, as Judd (1998) reminds us, was a growth in competition among American cities for investors, residents and labour. Large infrastructure and real estate development projects became the policy tools of choice in cities across the United States.

In the case of Brooklyn, for example, civic boosters began extensive planning and coalition building efforts in the early 1980s targeted towards the rebuilding of infrastructure in the downtown, with the goal of attracting and retaining commercial, retail and corporate tenants. These efforts resulted in a number of large-scale development projects including the redevelopment of Brooklyn's downtown commercial core at Fulton Mall, and the building of MetroTech, a high tech office development centre (Gross and Rogowsky 2000). The assumption was that these investments would enhance the local business environment and support more intensive use by residents. Over time, it was argued, the area would inevitably emerge in an organic way to support a more active tourist market; support the locality and tourists would come too.

Not surprisingly, as Gladstone and Fainstein point out, over the next three decades: 'In New York City the Koch, Dinkins and Giuliani administrations . . . worked closely with property development interests, relying on real estate development as the engine of growth' (Gladstone and Fainstein 2001, p. 34). Tourism promotion was addressed within a 'framework' of mass tourism. The goal was straightforward, developing and marketing places in an effort to maximise the volume of visitors – domestic, international, leisure and business. Tangentially, in light of perceptions of the City as dangerous, these developments were also often designed with a more fortress-like mentality, separating the users from the urban fray. The World Trade Center, for example, was designed to allow the visitor to arrive via commuter rail. The visitor was then deposited inside a complex of commercial, shopping and business offices. One never needed to leave the defined zone, never needed to step foot on the streets of lower Manhattan.

Lost from this approach was a real understanding or consideration of the contours of the 'demand' side of New York City tourism. For whom, then, was the City actually being developed? The business interests situated in the central City, in this case the area located south of 96th street on the Island of Manhattan, became both author and architect, albeit indirectly, of tourism development. Because the bottom line for these interests was fiscal gain, a.k.a. profits, development was increasingly targeted not at all tourists, but rather at select tourists. As Peter Eisinger suggests, while volume may be a desirable outcome, a more narrowly defined consumer market conditions this goal. Cities do not want to attract any tourist, they want to attract those who will spend money:

> Today . . . the city as a place is manifestly built for the middle class, who can afford to attend professional sporting events, eat in new outdoor cafés, attend trade and professional conventions, shop in festival malls, and patronize high- and midbrow arts.
>
> (Eisinger 2000, p. 317)

While development at the centre continued to isolate and selectively respond to the needs of some tourists, others, the domestic working-class visitor and ethnic visitors, and some of the international visitors in search of the 'real' City, sought out more distant venues outside the centre on the periphery of Manhattan, in places like Downtown Brooklyn, Long Island City Queens and Harlem in upper Manhattan. In these areas, communities began to organise themselves in an effort to capitalise on the emerging demand, seeking to expose the life of the neighbourhood rather than to enclose it.

This led to the emergence of New York's dual tourism market. The centre was increasingly being developed for the suburban, white middle-class, and business visitor, while new tourism areas (NTA) began to emerge from community-based efforts designed to serve working-class, ethnic and international tourists. In most cases, these NTAs developed gradually and organically – out of the mix of local history, art and culture. They were not planned as tourist areas. In fact it was their uniqueness as 'lived in' or 'worked in' neighbourhoods, that made them a draw for the 'new tourist', a person who seeks to experience the life of the City, and to mingle with those living in the City, rather than being separated from it.

NEW YORK'S TOURISM MARKET

While many cities vie for positions within the global tourism market, New York has remained a centre for visitors from all parts of the world. New

York has continually been ranked the top tourist destination in the United States, and is among the top ten internationally. Even in the aftermath of 11th September 2001, tourism was able to rebound within eighteen months of the terrorist attacks. In 2006, New York was ranked sixth in the world by the number of tourist arrivals, second by the number of air passengers, and tenth by the number of conventions held within its boundaries (Bremner 2007).

> The cheap dollar has also given the [New York] tourism industry a boost. The number of visitors set a record of 46 million in 2007, driven by a 17% jump in the number of foreign tourists, which contributed to strong sales tax collections. Hotel occupancy exceeded 86% and room rates averaged $304 – also a record. With the dollar expected to remain weak in 2008, tourism will remain strong.
> (New York State Office of the State Comptroller 2008, p. 3)

This is perhaps not surprising. NYC has substantial assets: roughly, 72,000 hotel rooms, 19,000 restaurants, 6,200 buses, 660 miles of subway track, almost 13,000 taxis (NYC & Company 2008), 1,500 weekly Broadway theatrical performances (The Broadway League 2008), 1,900 chartered museum and heritage organisations (Baldwin 2006, p. 20) and 13,000 buildings, including iconic structures like the Empire State Building, the Chrysler Building, the Statue of Liberty and Brooklyn Bridge. New York has been ranked second (behind Hong Kong) in terms of the visual impact of its skyline (Emporis 2008).

As Table 2.1 illustrates, almost forty-five million people visited New York in 2007. In sheer volume, domestic visitors, coming from the northeast of the United States dominate. These tourists tend to be predominately middle-class and suburban. Interestingly, while this may be the desirable market from the perspective of developers, they are not always the most desirable from the perspective of those in the tourism industry. Because the vast majority are day-trippers, their economic share of the market is relatively small – representing less than 30 percent of total tourist expenditure. The international traveller to New York spends almost four times the amount spent by domestic travellers per day. While domestic visitors spend on average one or two days visiting, international visitors tend to spend one week or more (Genoves *et al.* 2004).

For the tourism industry therefore, it is the international visitors that are the most desirable, because they dominate in relationship to fiscal impacts. While only representing 10 percent of the total number of visitors to New York, they represent 40 percent of the expenditures (Bram 1995, p. 1). Foreign tourist markets therefore, must be viewed as critical. Even during domestic recessions, the foreign tourist is likely to continue to visit (Clendinen 1980). A weak dollar is advantageous for the foreign tourist, thus, as early as 1980, New York's tourism marketing programme was expanded and the 'I Love

Table 2.1 Composition of New York's tourism market: Number of visitors 1998–2006

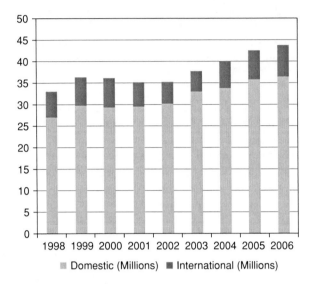

Source: New York City and Company (2008).

New York' campaign went international. The result, as Table 2.2 below reveals, is an impressive and ongoing dominance by New York of the international tourism market in the United States.

In raw volume, New York drew more that double the number of international visitors to its neighbourhoods than all other cities in the United States (U.S. Department of Commerce International Trade Administration, Office of Travel and Tourism Industries 2008).

Table 2.2 New York City ranking and market share of international tourism in the United States

Year	Rank	Market Share
2006	1	28.7%
2005	1	26.8%
2004	1	25.4%
2003	1	22.1%
2002	1	22.2%
2001	1	22.0%
2000	1	22.0%
1999	1	22.5%
1998	1	21.1%

Source: U.S. Department of Commerce, Office of Travel and Tourism Industries (2008)

While most think of New York through the lens of Manhattan, it is important to recognise that New York is made up of five boroughs each with its
own distinct identity and interests. The result is what I would suggest is
the existence of a dual tourism market. Manhattan (south of 96th Street)
represents one market targeted towards a more upper income and business
clientele – under the aegis of 'New York as a World City'. The boroughs
of Brooklyn, Queens, the Bronx, Staten Island, and Upper Manhattan in
combination might be better understood as representing a second market
– characterised by efforts to promote neighbourhoods to domestic and
ethnic interests – under the aegis of a 'New York of Neighbourhoods'. *Dual
markets* lead to the emergence of *duel agendas*.

New York Tourism Politics: Dual Markets and Dueling Agendas

Tourism politics in New York is played out indirectly in political battles over
cultural resource allocation, economic development support, and in control
over land use planning. The geographic bias of tourism related policy support in New York is on the central City, which in New York translates to
investments made on the Isle of Manhattan south of 96th street. Marketing
focuses primarily on the major Manhattan cultural institutions that include
the Museum of Modern Art, the Metropolitan Museum of Arts and Lincoln
Center for Performing Arts. For example, a quick look at the NYC & Company web site, the official marketing and tourism organisation for New York
City, shows that the majority of attractions (over 70 percent), museums (over
67 percent), and restaurants (over 88 percent) listed are Manhattan-based.[2]
Similarly, public and private development efforts tend to focus on the
building of large multi purpose entertainment centres in the central City.
For example, the Times Square re-development, a two billion dollar project
described as 'the largest development effort the City and State have ever undertaken . . . a far-reaching attempt to turn 42nd street between Broadway and
Eighth Avenue into a . . . commercial/cultural thoroughfare for private
business and the public' (Bell 1998, p. 26). The more recent Time Warner
Center, was a $1.7 billion project, described as an 'iconic mixed-use property
developed by Related Companies and Apollo Real Estate Advisors, L.P., that
serves as the world headquarters for Time Warner Inc.' (Building Design &
Construction 2008). There is also almost continuous promotion of Broadway
Theatre. This Manhattan centric approach to development shaped the contours
of tourism promotion in New York for much of the twentieth-century.

While Manhattan has been the primary locus for public and private tourism
related investment, New York is in reality a regional city, uniting five coun-
ties (a.k.a. the boroughs of Brooklyn, the Bronx, Manhattan, Queens and Staten
Island) under one jurisdiction in a general-purpose local government. New
York City operates under a strong Mayor, weak council system. The Mayor
is the executive, with significant powers in setting overall budget levels, and

in land use planning and appoints the majority (chair and six members) of the twelve-member New York City Planning Commission. The New York City council is the legislative arm, whose fifty-one members, in principle, represent a significant counterweight. A two-thirds majority can override mayoral vetoes of legislation. Each borough has a borough president, whose powers were severely curtailed following the 1989 New York Charter revisions, when the New York Board of Estimates was eliminated – a primary agency for budget decisions. The most local level of government are the fifty-nine Community Boards (CB), each made up of roughly fifty appointees of the borough president and local city council members. The CBs offer advice on local budget needs, service allocations and hold public hearings on land use issues.

Differences in demography, economy, patterns of development and conditions of the built environment combine to produce a complex pattern of needs and demands. The result has been 'a volatile and often raucous' politics (Gross and Rogowsky 2000), in which politicians commonly ally with constituency and borough interests, rather than forming cross-borough coalitions – the result for tourism policy has been a series of 'dueling' tourism agendas. The duels are played out in annual battles for resources that not only pit boroughs against one another, but at times also promote divisions within each borough as each of New York City's fifty-nine community districts do battle with one another over their share of the fiscal pie.

The Manhattan tourism agenda reflects a 'world city' branding approach – in which mega projects, iconic architecture, and privately led–publicly supported development dominate. This approach tends to garner support from the Mayor, the eight Manhattan city council members representing constituencies south of 96th Street, the Manhattan Borough President and a variety of private sector interests. This is the agenda that dominates New York City tourism politics even though the boroughs of Brooklyn, the Bronx, Queens, and Staten Island accommodate the bulk of New York's population and account for the majority of its landmass and also control the majority of seats on the New York City Council (forty-seven out of fifty-nine members).

In the realm of tourism related policy and resource allocation, members tend to ally with borough and constituency interests, rather than supporting cross borough alliances. Competition, therefore, rather than collaboration is the common pattern outside of the Manhattan core. The ability for the Manhattan group to collaborate, while other political actors compete, supports the Manhattan-centric bias in tourism politics. Not surprisingly, the Manhattan cultural institutions receive more that double the amount of funding from the New York City Department of Cultural Affairs than Brooklyn, and more than ten times the amount received by Queens (New York City Office of Management and Budget 2007).

In turn, if for example the Brooklyn and Queens council representatives chose to form a single coalition (making up thirty-two of the fifty-nine members), they would, in principle, be able to exert great influence. The reality has been a situation in which council members fight for their local constituency first. If they collaborate, which is the exception and not the rule, it is most commonly in borough delegations (party is somewhat irrelevant in the New York City legislature, as the City Council is almost completely Democrat – there were only three Republican members in 2008). In turn, it is local interests, race, geography and class forces that shape New York's political agenda rather than party. These geopolitical divisions are mirrored in tourism politics, with the agendas of the Bronx, Queens and Staten Island being commonly driven by 'neighbourhood' branding efforts with the constant pull of global pressures tending to push these more locally driven agendas to the political periphery.

Brooklyn tends to be pulled in both directions, with the downtown business core tending to mirror Manhattan's world city branding efforts, and the interior tending to adopt the neighbourhood approach. In the political battles for resources, 'neighbourhood branding' is often subordinated to 'world city' branding efforts. As the director of one cultural institution in the borough of Queens commented:

> 'You have to work twice as hard as a Manhattan institution,' said Rochelle Slovin, executive director of the American Museum of the Moving Image in Long Island City, a film and broadcasting museum that is one of the most successful fund-raisers outside Manhattan. 'But we've all learned to live within these constraints. It's not fair, but it's a fact of life.'
>
> (Firestone 1994)

This enduring intra-city 'duel' for equity in resource allocation and public policies is critical in building an understanding of tourism politics in New York.

It was in light of this lack of political support from the central City that the boroughs outside of Manhattan began to explore new more locally driven and controlled economic development strategies and neighbourhood branding efforts. Here, communities built upon existing cultural and historic assets, in an effort to complement existing local activities, while simultaneously 'valoriz[ing] multiculturalism and diversity, giving rise to new forms of cultural capital and creating interest in formerly unattractive places' (Hoffman 2003a, p. 286).

Brooklyn politicians were early adopters of an outer borough tourism strategy. In the late 1970s, Brooklyn Borough President, Howard Golden set up a non-profit organisation called the 'Fund for the Borough of Brooklyn', to promote the borough's cultural resources and use the arts as a driver of neighbourhood economic development (Lichtman 2006). Then in 1979, the 'Fund' was replaced by BRIC (Brooklyn Information and Culture), a multi non-profit project that oversaw a variety of activities in the borough.

Perhaps the most well known was 'Celebrate Brooklyn', one of New York City's longest running, free outdoor performance arts festivals, started in 1979 in an effort to overcome negative perceptions of the borough's central park – Prospect Park.

These early initiatives, although tourism related, were far more focused on promoting the area to the broader metropolitan community. Thus Brooklyn turned its gaze upon Manhattan and the surrounding suburb of Long Island, and developed strategies to attract these users. During the 1970s, the areas and its attractions were marketed to domestic visitors, affluent local users and commercial development interests. During the late 1980s, efforts were made to rebuild infrastructure, attract and retain commercial, retail and corporate customers.

Success in and around Downtown Brooklyn and Prospect Park created some internal conflicts over the direction of the Borough's tourism vision. Downtown Brooklyn's began to mirror the Manhattan development and business led, world city branding efforts – trying to draw in corporate interests as a driver for development, while ethnic communities such as Fort Green and Williamsburg, areas that lacked major cultural institutions, developed their own locally based coalitions and institutions. Brooklyn tourism politics was divided – dueling agendas. In 1994, the borough president sought to overcome the divisions and provided funding for the creation of the Brooklyn Tourism Council.

The Queens Council for the Arts might be credited for spurring the contemporary tourism efforts in the Borough of Queens. Started in 1966 as an organisation to support the arts and artists, they spearheaded the efforts to promote and market the Borough's ethnic cultures, through the designation of the No. 7 train line as a National Millennium Trail in 1999. An effort that took over five years to achieve, due in part to the absence of a cross-borough tourism coalition to provide unity and support for the action. In 2003, the Queens borough president, Helen Marshall, invited a wide range of actors to come together to design a tourism strategy for the borough. The core conclusions were that the borough should market itself first to its own residents, and then others in the larger metropolitan region. From this initial meeting, a Queens Tourism Council emerged made up of 19 cultural organizations, five civic organisations and one hotel (Robinson 2004). The make up was quite different from the Brooklyn Tourism Council that had a much larger corporate and commercial presence.

In Harlem, tourism is far more bifurcated. In the case of Harlem (Manhattan North of 96th Street) a tension exists between local and global branding efforts. Its proximity to the Manhattan core has resulted in its exploitation by some central Manhattan based tourism interests. Although some 800,000 people were reported to have visited Harlem in 2000 (Hoffman 2003b, p. 94), the vast majority did so within the confines of tour buses – affectionately known as 'drive-by-tourism' in New York. The tour bus operators – situated in the central City – were the most direct beneficiaries. In the neighbourhoods only select destinations, like the Apollo Theatre, reaped any direct benefit, and always with a fee paid to the tour operators. Tourism grew throughout the last decade yet few in the community reaped any direct profit. In Harlem, neighbourhood coalitions have tended to evolve in an effort to preserve and support local interests, and to ensure the community is able to benefit from this emerging area of economic development. Multiple organisations exist, and each pursues their own agendas. The Upper Manhattan Empowerment Zone has a tourism development strategy. The East Harlem Board of Tourism, is a non-profit organisation made up of arts and cultural groups in the area. Harlem One Stop is another cultural tourism group located in West Harlem. Thus, here a more divided political environment might be the best descriptor for this area's tourism politics, but with 'neighbourhood' rather than 'world city' branding at the core of the local agenda.

The separate efforts of each borough were then given a boost in 2001, when the Ford Foundation spearheaded a citywide initiative to promote community tourism in New York neighbourhoods, outside the central tourist attractions. Tourism coalitions were formed to promote neighbourhoods (Robinson 2004). In each case, communities began to strategise around the effort to promote existing culture and heritage in these non-traditional tourist destinations.

While the boroughs developed these strategies, they often did so in distinct and different ways. All incorporated community-oriented approaches, putting neighbourhood branding at the centre with variable degrees of responsiveness to global interests. Brooklyn has tended to straddle the global/local dynamic – often pulled in both directions. In the case of the Downtown, tourism actors have sought to meld these visions into one. Queens and Harlem have adopted more locally focused political strategies, choosing to market themselves first to local residents and second to visitors. The divisions between, and within, the boroughs creates 'dueling agendas' that often leave these communities short changed in resources and development support.

NEW YORK'S 'OFF THE BEATEN TRACK' NEW TOURISM AREAS

While politicians in this world city may lack the will to provide equal fiscal support for a outer borough tourism, this does not seem to be the case for the tourists themselves. The neighbourhood branding efforts have created an interesting niche for the more adventurous traveller seeking out more distant locations beyond the central City. This research on tourism demand suggests that new tourism areas represent a critical component of world city tourism, and this is even more important given the homogenising trends shaping tourism development in the central City. We will first describe these areas and then discuss the tourists they seem to be attracting.

The Brooklyn NTA and its Environs

> Ten years ago, tourists visiting Brooklyn saw it as nothing more than a side trip from Manhattan. Maybe they walked across the Brooklyn Bridge or took the subway out to Coney Island, but few had dinner or stayed the night. These days, however, Brooklyn is a destination unto itself. Now visitors to Brooklyn 'stay here and go into Manhattan for the day – or they don't go to Manhattan at all', says Monique Greenwood, who runs the Akwaaba Mansion, a B&B in Bedford-Stuyvesant. 'Most of my European guests have already done Manhattan. Now they want to see Brooklyn. They're going to the Brooklyn Museum, the Botanic Garden, the Brooklyn Academy of Music, and the Brooklyn Heights Promenade. Or they're just hanging out in Brooklyn. They like the idea that it's more of a neighborhood here. They believe that Brooklyn is the hip borough.'
>
> (Harpaz 2006)

Brooklyn is the largest of the City's five boroughs by population. With almost 2.5 million inhabitants, it would be the fourth largest city in the United States if it were independent. It has experienced a 7.2 percent increase in population over the past decade. Brooklyn occupies 81.8 square miles of land, of which 38 percent is exclusively residential, 4 percent is mixed use, 3 percent is commercial, 5 percent is industrial, and almost 34 percent is open or recreational space (New York City Department of City Planning 2006a).

As a tourist destination, one can also look at the Brooklyn environs through the lens of popular culture. The area is commonly portrayed as working-class, as seen in classic television shows like the 'Honeymooners'; is depicted as ethnic Italian in 'Moonstruck' and 'Saturday Night Fever'; in Spike Lee films, like 'Do the Right Thing', and 'Smoke', or the more recent television show by Chris Rock 'Everybody Hates Chris' we experience a racial vision of Brooklyn and it is also portrayed as dangerous, as in films like 'Girl Fight', 'He Got Game', and classics like 'Dog Day Afternoon'.

> [And] for years, countless movie characters identified as coming from Brooklyn spoke only the King's English. The Kings County English, that is, laced with puhlenty of dese, dems, dose and de like. The mere mention of Brooklyn was good for a screen laugh, since everyone knew that only a slap-happy bunch of lovable mugs lived there.
>
> (Haberman 2003)

Clearly, Brooklyn has a long pedigree in popular culture – which is a mix of eccentricity diversity, danger and of course the Dodgers (Brooklyn's lost baseball empire).

Brooklyn is located in the south-eastern portion of New York City, and is attached to Manhattan by three bridges – the Brooklyn, Manhattan and

Williamsburg – and one tunnel – the Brooklyn Battery. It has a vibrant community life, shopping for all types, art, music, a wide variety of eating options, and is easily accessible to Manhattan.

With the opening of the Brooklyn Bridge (the northern tip of our study area) in the late nineteenth century, western and central Brooklyn neighbourhoods rapidly developed. The area between the Brooklyn Bridge and Prospect Park contains many neighbourhoods that each has their own offerings to their communities and visitors. This area includes some of the most beautiful homes and historic districts in New York City. This new tourism area crosses three community districts, and incorporates a range of cultural attractions. The area is very ethnically diverse with African-American, Latino and White populations dominating depending on the precise area (New York City Department of City Planning 2006b and 2006c). In addition to its historically preserved brownstone neighbourhoods, the area also includes Brooklyn's civic (Borough Hall) and commercial centres (Metrotech and Fulton Mall). While the area is resource rich in tourist attractions, scattered locations presents a major challenge to building a cohesive NTA.

The NTA study area begins at the foot of the Brooklyn Bridge (just across the East River from New York's City Hall and Lower Manhattan) and extends into central Brooklyn, ending at the far end of the Brooklyn Botanical Gardens and Prospect Park. It is a somewhat more established tourist destination than Western Queens (one of the other NTA study areas) because it incorporates a variety of long standing cultural institutions and historic sites. It includes Prospect Park (designed by Frederick Law Olmsted and Calvert Vaux, and founded in 1867), the Brooklyn Museum (founded in 1897, and home to the second largest art collection in the North East) and the Brooklyn Botanical Gardens (founded in 1910) and attractions like the Brooklyn Academy of Music (the oldest performing arts venue in the United States founded in 1861). The Brooklyn NTA study area is best described as an area that has evolved over time, marketing its distinct cultural and historic areas, yet it suffers from both the geographic separation of its attractions across multiple neighbourhoods, and from perceptions of the area as dangerous.

The existence of these historic and well-established institutions helps to explain both the area's early organising efforts in the 1980s, and its early adoption of a real estate and commercially led tourism agenda as a vehicle to grapple with recessionary economic decline. Membership in Brooklyn Information and Culture (BRIC), and the Brooklyn Tourism Council were weighted far more heavily towards local business and commercial interests than to community residents. In addition, the long standing cultural institutions had been informally allied with one another for nearly a century, thus their voices and interests often dominated. Because the NTA programme mimicked elements of the Manhattan approach, had long-standing cultural institutions, and was pulled between alliances with Manhattan and with its neighbouring constituencies in Brooklyn, downtown Brooklyn politicians commonly found themselves walking a tightrope between global and local development pressures. A situation that continues today, as seen in the battles over the development of Atlantic Terminal (at the centre of the NTA) in Forest City and Ratner's large scale commercial and residential development project, anchored by the iconic architecture of Frank Ghery, in the form of a new basketball arena for the Nets basketball team.

The neighbourhoods surrounding the NTA are, for the most part, very residential. With small 'mom and pop' shops, new ethnic venues like 'Bogolon', an emerging African-America craft hub, restaurants, and bars, this neighbourhood was not only used by tourists, but also by residents of Brooklyn and greater New York City.

The Queens NTA and its Environs

Queens is the largest borough by landmass, spanning some 112.2 square miles (35 percent of the New York City). It is also the fastest growing borough, its population increased by 14.2 percent over the last decade. At the last census, the population had reached almost 2.3 million. While the borough as a whole is over 46 percent residential (largely dominated by one and two family homes), 2 percent mixed use, 4 percent industrial, almost 4 percent commercial, and 19 percent is open space/recreational (New York City Department of City Planning 2006d). Long Island City, the location of our NTA is one of the most industrial portions of the borough with over 11 percent of the community district being identified as industrial (New York City Department of City Planning 2006e). If we add to this the extensive land taken up in this district for transportation (almost 8 percent), it becomes clear that Long Island City has a very different feel to it that central and downtown Brooklyn.

An interesting feature of Queens, as distinct from Brooklyn, is its difficulty in establishing an identifiable and unique borough wide identity. This may explain the strategy of Queens' politicians to work collaboratively across communities. The Queens' delegation at the City Council, for example, has often been able to exert great political power due to their efforts to promote a unified front, working first to market itself to its local residents and branding itself through its local identities as opposed to Brooklyn's more global/local approach.

> The establishment of Queens as a municipal borough [in 1898] did not create, out of the blue, a Queens identity. . . . the area had many centers and diverse local traditions, and much of it still existed as farmland and open terrain. Furthermore, the new Queens County, which was made coterminous with the borough, had been severed from its former eastern territory, which had become Nassau County. The new Queens, in short, was a political artifact without a real past in the life of the people.
>
> (Seyfried and Peterson n.d.)

Thus, Queens is much more divided as a borough and identity tends to attach to neighbourhood, rather than the borough as a whole.

Perhaps the most widely known, and most damaging, portrayal of Queens came from the American sitcom 'All in the Family'. Archie Bunker was the central focus of this as the bigoted, white conservative who was known as the 'equal opportunity offender'. Despite this negative portrayal, that suggested a homogenous community unaccepting of diversity, almost 48 percent of the borough's population today is immigrant and it has a long history of mixed ethnic communities. Lesser-known facts about the borough include the fact that it was the home to some of New York's most famous jazz musicians, including Louis Armstrong, Count Bassie and Ella Fitzgerald.

Long Island City sits in the south-western portion of the borough and is connected to Manhattan (just east of midtown) by the Queensborough (a.k.a. the 59th Street) Bridge. Long Island City sits within Queens Community District 1, and abuts Astoria, in District 1, and Hunters Point, Sunnyside and Woodside, in District 2. The area is 42 percent white, 27 percent Latino, 13 percent Asian, and 10 percent African-American (New York City Department of City Planning 2006e).

If the Brooklyn NTA was a primarily residential area, when we come to Long Island City, we are presented with a very different experience. Because the area was primarily industrial, it naturally lent itself to reuse by local artists. As such, while our Brooklyn NTA was largely a historic cultural and neighbourhood area, in Long Island City we find an artist's enclave operating alongside a small working light industrial area.

> From the Manhattan side of the East River, Long Island City is a collection of motley landmarks – the towering Citibank tower; the candy-cane smokestacks of a massive power plant; the red-lettered sign of Silvercup Studios, where 'Sex and the City' and 'The Sopranos' are filmed. But there's much more than meets the eye to this well-established industrial area, the largest neighborhood in the New York borough of Queens. Nestled among its factories, foundries and warehouses are hundreds of artists' studios and some of New York's most notable arts institutions.
>
> (Mehta 2004)

While Central and Downtown Brooklyn included a somewhat bucolic neighbourhood escape, Long Island City might be better characterised as an area of urban 'edge'. Not surprisingly, it is estimated that some 2,000 artists work here. A composite of factories and waterfront, this area has had a more difficult time garnering support for its tourism, lacking the architectural lure of both Brooklyn and Harlem, and possessing a grittiness that makes it feel a more dangerous location.

The area has slowly been converted to an off the beaten track arts community – with the PS1 contemporary art centre (housed in a refurbished school house built in 1893); Socrates Sculpture Park (the only park in the New York area devoted to large scale sculpture exhibitions); the Noguchi Garden Museum housed in a former photoengraving plant, and of course, Silvercup studios where the 'Sopranos' and 'Sex and the City' were filmed. Within the same community district is also New York's largest prison – Rikers Island.

The Harlem NTA and its Environs

Unlike our other two study areas, Harlem is located just north and slightly east of central park on the Island of Manhattan. Harlem spans two community districts – 10 (Central Harlem) and 11 (East Harlem). This area covers a smaller geographic area than our other two areas (just 3.6 square miles). While 43 percent of the central Harlem area is residential, unlike the Queens' case, this is a far denser community with only 3 percent being one or two family homes, and some 42 percent being multifamily residential. Here there is also very little industrial use, less than 1 percent average across both community districts, 5 percent is commercial and 5 percent is mixed use. Seventy-seven percent of the Harlem population is African-American, 17 percent is Latino – and 2 percent is White (New York City Department of City Planning 2006f and 2006g). As such it offers a very different ethnic mix than either of our study areas in Queens or Brooklyn.

Harlem also has its own unique cultural identity shaped by both a romanticised vision of the 1920s Harlem Renaissance, racial conflict and its post World War Two history of concentrated inner-city poverty. It is also home to a variety of cultural venues, such as the Apollo Theatre and the Cotton Club, as well as being a centre of religion with the Abyssinian Baptist Church and St. John the Divine being but two of the 400 plus religious venues housed here.

In popular culture Harlem is most commonly portrayed as a centre of racial conflict and cultural renaissance. The early history of the area is portrayed in films like the 'Cotton Club'. Films like 'Cotton Comes to Harlem', described as part of the 'blaxploitation' genre, portrayed the area and the ways its population were exploited by white America. 'Carlitos Way', however, takes us into the seamy side of East Harlem's gangsters and crime, while the cult classic, 'The Brother from Another Planet', drops an alien (in the form of a homeless black man) onto the streets of Harlem. As one travel writer asserts:

> Yes, Harlem has its sketchy stretches (mostly further north and east) but it also has a kind of faded, almost Venetian grandeur. The wide streets are lined with once opulent apartment buildings and rows of brownstones among which the newly restored stand out beside the derelict, like flowers growing amid stinging nettles.
>
> (Souter 1997)

The area is a mix of musical venues, ethnic restaurants, historic buildings, brownstone neighbourhoods, museums and churches. Well connected to the centre via multiple subway and rail lines, the area is easy to access.

Non-Traditional Tourists

It was speculated that the contours of each of these areas would likely produce differences both in the kinds of tourists visiting each area, and that overall those visiting NTAs would also be different to the tourists visiting the central City. Of the three areas, the Queens and Harlem locations were advantaged geographically due to easy transportation links and a concentration of attractions in a relatively small area that was easily navigable. Brooklyn and Harlem had cultural and historic advantages. Both areas have identifiable cultural monikers, while the Queens NTA lacked a distinct identity and had done little to market its history. Thus, the Queens NTA was shaped primarily by its collectivity of arts venues, but remained somewhat anonymous as a destination.

So, who is visiting these NTAs? Are all NTAs equal in the eyes of the visitor and, to what extent do we find the presence of a 'dual' tourism market? Are those visiting these areas representative of the tourists who visit the central City? A total of 219 tourists were interviewed between 2004 and 2005 of which almost 43 percent were international and 57 percent were domestic. A much higher concentration of foreign tourists were visiting these NTAs as compared with the overall numbers visiting the central City (NYC & Company 2003). Of the three NTAs, only Brooklyn had a higher concentration of domestic than foreign visitors.

When probed about the type of accommodation used, of those visiting our NTAs over 60 percent indicated that they were staying with friends or family, and the reminder were staying either in hotels, bed and breakfasts or other rented apartments. In contrast, according to NYC & Company, the majority of those visiting the central City stay in hotels.

Most of the international visitors came from West Europe. However, there were small differences in the composition of other foreign visitors in each area. In the Brooklyn NTA, after West Europeans, the second largest group (22 percent) came from Pacific Asia, followed by South and Central America (17 percent) and Canada (8 percent). In Harlem, 73 percent of visitors came from West Europe, followed by small percentages of tourists from Canada (11 percent), Australia (5 percent) and East Europe (5 percent). Long Island City, like Harlem, had most tourists coming from West Europe (70 percent), and small groups from Australia (10 percent) South America (5 percent), Pacific Asia (5 percent), the Middle East (5 percent) and Canada (5 percent).

Tourists visiting NTAs were themselves urban denizens, across all three NTAs the majority indicated that they themselves came from large cities.

This suggests that those most likely to step off the beaten-track are also those who are comfortable in urban areas.

Some 80 percent of those interviewed indicated that they had arranged their visits on their own, and almost 60 percent said that the Internet had been especially useful in planning their trips. By way of contrast, NYC & Company (the tourism authority for the City) found that 55 percent of those visiting the central City relied on a travel agent. This again suggests an interesting contrast – non-traditional tourists, also utilize non-traditional sources of information. This may also suggest that travel agents are not touting NTAs as heavily as they do the central City.

We also sought to understand why tourists were visiting New York. Thirty-six percent indicated that they were in New York to visit friends or relatives, 35 percent indicated they were in New York specifically for leisure, 10 percent for business, 6 percent were travelling around the United States and New York was but one of many stops, 5 percent were here for study and 4 percent were in New York for a social event. This points to the importance of leisure based tourism while, in the central City, there is a larger mix with business tourists as well.

Visitors to NTAs were generally not first time visitors, rather most had been to New York three or more times. This suggests that repeat visitors are more likely to seek out an NTA experience. At the same time, there were differences found between our study areas. The Queens NTA attracted a larger proportion of new visitors while Brooklyn attracted a larger number of repeat tourists and Harlem tourists were more equally divided between the new and the old. As Brooklyn was the area with the most difficult geography to navigate, it is not surprising that new visitors tend not to venture here.

We felt it important to explore the degree to which tourists visiting NTAs were venturing beyond specific destinations and experiencing the area. Here we found clear differences between neighbourhoods. Geographic concentration appeared to be an advantage for both Harlem and Long Island City, but a disadvantage for Brooklyn. Some 43 percent of visitors to Harlem indicated that they planned on dining in the area, as compared with 23 percent in Long Island City and 18 percent in Brooklyn. Likewise, larger proportions of new tourists intended to shop or just explore the areas. Brooklyn consistently had fewer tourists who were comfortable with exploring outside of their specific venues. When asked, the Brooklyn tourists indicated that they did not know where to go, and remained fearful of exploring. Visitors to Long Island City enjoyed the artistic vibe, but felt the area felt very isolated and deserted – there was little activity on the streets. In Brooklyn, visitors pointed to the architecture, the greenery and the people as what they most liked, but felt that many of the tourist venues were inaccessible. In Harlem, people liked the ethnic diversity, convenience and the history, but many remained fearful of what they described as very busy streets that felt dangerous.

All those visiting these areas indicated that they would visit the area again – this was true of 77 percent of those visiting Harlem and Brooklyn, and 65 percent of the Long Island City tourists. Over 80 percent in each neighbourhood said that they would recommend the area to their friends.

CONCLUSIONS

So what does this tell us about New Tourism Areas and the tourists they attract? First, one cannot help but notice a real difference in the approach being taken to tourism outside the centre. Rather than making major investments in mega projects, the outer boroughs are instead making efforts to expose existing assets – ethnic culture, art, history and architecture. These areas are marketing 'neighbourhoods' and attempt, as is common across Europe, to integrate residents, tourists and workers. Promoting what many would consider a more 'authentic' urban experience, in which the visitor is not separated, but rather integrated, into the life of the City.

The biggest barrier faced in each of these areas seemed to be perception – fear of the unknown and acceptance of the cultural depictions of areas as dangerous remain significant barriers to the development of NTAs.

The new tourists that we found were clearly seeking a more integrated tourist experience – unifying culture, history and community. They might be described as more sophisticated, individual and well-informed types. They were seeking a more authentic New York experience, rather than the traditional, or manufactured, tourist experience. The tourists planning longer visits were also more likely to visit NTAs and, as mentioned, the new tourists were also more likely to visit NTAs if they were returning visitors (over half of the surveyed visitors had visited New York more than three times).

NTAs are a growing trend in urban vacationing and the 'new tourist', will not be new for long, but will eventually become a more typical visitor to cities. More and more, people are looking for a distinctive, and non-homogenous, experience and with the advent of globalisation, tourists have to look to the NTAs to find the heart of the cities they visit.

At the same time, as these areas lure more tourists, they risk being homogenised in much the same way as the central City. Evidence that this type of trend is crossing the East River to Brooklyn is growing – the building of the Brooklyn Nets Arena that requires the tearing down of some 150 Brooklyn brownstones, just down the block from the Brooklyn Academy of Music, points to the growing influence of corporate, over community, based interests in tourism promotion.

Thus, the success of these outer borough NTA efforts, as tourism venues, is also problematic. Despite the growing focus by inner City communities on developing their own tourism markets, they will be faced with growing pressures from business interests seeking to capitalise on their gains, and in some instance to exploit these locally developed ethnic enclaves.

One indication of a changing pattern came in 2004 when NYC & Company (the City's tourism bureau) held its annual luncheon in Brooklyn, 'the first time in at least 25 years that the City's tourism agency had held such an event outside of Manhattan' (Kuntzman 2004). A second more significant indication of change came in 2006, when Mayor Bloomberg promoted a 'five borough' economic development strategy, in which tourism was earmarked to play a more prominent role in local development efforts (New York City Mayor's Office of Industrial and Manufacturing Business 2006).

Thus, as we consider the future of New York as a World Tourism City, we can see that the image of the world city is different from the city of neighbourhoods – it is unclear whether the dual city will continue to assert itself or if the world city will slowly overtake the other.

3 Tourists, urban projects and spaces of consumption in Paris and Ile-de-France

Patrizia Ingallina and Jungyoon Park

INTRODUCTION

In this chapter, we examine the development of tourism in Paris not just through its tourism figures, but through the broader planning strategies and tourism policies that are changing the way in which the City thinks about its attractiveness in global competition. We will see how new attractive places in Paris, that are often expected to provide sustainable and alternative forms of leisure and tourism, can generate urban, social, and economic benefits for local communities and provide more creative tourist experiences for visitors. In the latter part of the chapter, we examine the case of Bercy Village, created in 2001 as a consumption space for residents and visitors alike. In order to make sense of this new approach to tourism in the City, we need to understand broad trends and concerns about the continued attractiveness of Paris to outsiders, and concerns about the City's ability to hold on to its middle-class residents. In addition to looking at the context of tourism trends and emerging policies, we also need to see changing public policy in a wider context, a shift in attitudes to urban development and planning, and the role of the 'urban project' in the transformation of French cities. In this way, we can understand the distinctiveness of the Paris case as the City aims to change its image and remain competitive in the global competition for visitors and global competition between the major centres of business wealth and heritage and culture.

In this context, we start by examining the development of tourism in France and the Paris region.[1] This section of the chapter reviews the dominant general trends, and national and regional tourism policies, and examines developments in the tourism policies of the Paris Region. We then move on to consider how urban competitiveness – competition for visitors, for economic growth, and to attract and retain a middle-class – has been conceived in the French context. As the spaces of the City are redeveloped and places remade, particularly important is the concept of 'urban project' and the relationship between formal and informal design tools and notions of territory and territorial attractiveness. In Paris and elsewhere *politiques de projet* are important in re-branding the City and promoting the image of particular places.

This discussion of French urban planning is then related specifically to the case of Paris and Ile-de-France (see Figure 3.1).

The third part of the chapter examines some new developments in the City's approach to tourism. We look briefly at regional developments across Ile-de-France and, in Paris, at the quarters of Belleville and the Viaduc des Arts, recently promoted as new tourist locations, and how they contribute to the re-branding strategies of the City New tourist paths in Belleville are enthusiastically promoted by civil associations and, little by little, have engaged the support of the City Council. We examine in greater detail the case of Bercy Village and its meaning in the broader urban strategy for the development of eastern Paris. New retail development in Bercy Village and along the Viaduc des Arts, originally designed for the local population, attract more and more visitors from other regions and from overseas. We consider how local authorities handle their urban planning for these sites and how they can respond to the search for new images of the City through tourism and city marketing.

At the heart of the chapter are questions concerning the notion of French 'urban project' and about how the Paris region views its competitive position and attractiveness in the context of the increasing global competition. We ask how local authorities with different territorial scales are developing a greater awareness in public policy of the urban preferences of residents and of visitors. The chapter responds to a series of questions: How are urban

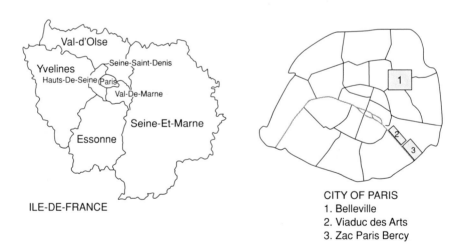

CITY OF PARIS
1. Belleville
2. Viaduc des Arts
3. Zac Paris Bercy

ILE-DE-FRANCE

Ile-de-France region is one of 22 French metropolitan regions with elected government. The region includes the city of Paris which has its own mayoral government and administration and about 1200 other Communes grouped into seven Départements. In many places we use the shorthand 'Paris region' to refer to Ile-de-France.

Figure 3.1 Paris and Ile-de-France
Source: Adapted from www.iledefrance.fr

policies and city marketing approaches changing the vision of a city and its spaces (Ingallina and Park 2005)? How are these questions reflected in policies towards tourism? How can Paris maintain and increase its new tourist attractions? What are the main concerns in these recent strategies in tourism and what conflicts may be generated by the new strategies for tourism development?

TOURISM IN FRANCE AND THE PARIS REGION: SOME KEY RESULTS

Economic weight of tourism in France

In 2005, income from tourism represented 6.4 percent of GDP (108.1 billion Euros) and constituted one of the most important sectors of the French economy. This amount is made up of the spending of French tourists (55.5 percent) and foreign tourists (35 percent) who numbered 76 million and the domestic spending of French tourists preparing for travel to foreign countries (9.5 percent) (French Ministry of Tourism 2006b). The number of foreign tourist arrivals increased to 79.1 million in 2006 and, in the same year, there were 826.6 million overnight stays made by French residents and 79.1 million international arrivals who made 119.5 millions of day visits and 497 millions of overnight stays (French Ministry of Tourism 2007). According to a survey of the French Ministry of Tourism in 2007 (French Ministry of Tourism 2007), France is still the first country in the world in attracting foreign visitors and, until 2001, French tourism had an excellent level of growth even if the growth has stagnated somewhat in recent years. Recently, because of the ever-increasing worldwide competition to attract tourists, France has had to share the world tourist market with many new rising tourist destinations (see Table 3.1).

Tourism can generate up to two million jobs during the high season (both direct and indirect employment) and those jobs are distributed among 232,000 businesses. Most of these are small and medium sized companies (French Ministry of Tourism 2006c). Annually, 894,000 wage earners are directly linked to the tourist activity (direct employment), rising from 686,000 in January to a maximum of at least 1,184,000 in August. In 2005, hotels, cafés and restaurants employed 828,200 workers and we can also add about 170,000 self-employed workers in this sector. There were 15,700

Table 3.1 Evolution of the share of tourism in GDP from 2000 to 2005

Year	2000	2001	2002	2003	2004	2005
Tourism/PIB	6.8%	6.6%	6.6%	6.5%	6.5%	6.4%

Sources: INSEE / Account of Tourism 2006

salaried employees more in this sector in 2005 than in 2004 (French Ministry of Tourism 2006a).

In terms of the balance of earnings from international tourism, in 2005 foreign tourists in France spent about 34 billion Euros. This represents a rise of 3.5 percent compared to 2004, while the spending by French tourists abroad reached 25.1 billion Euros (an increase by 8.9 percent). Even if we can see a substantial surplus of 8.9 billion Euros compared to 2004, we have also to take note of a decrease in this balance by 9.2 percent resulting from the growth of French people's overseas travel (French Ministry of Tourism 2006a). Nevertheless, this balance is more positive than that of the car industry (8.5 billions) and the agro-alimentary sector (6.3 billions), demonstrating the weight of the tourism industry in French foreign trade (French Ministry of Tourism 2006b).

Within France, Paris and the Paris region compete with the region of Provence-Alpes-Côte d'Azur (PACA). If we compare the tourism performance of French regions, in 2005 PACA had 13.9 percent of French overnight stays (overseas visitors accounting for 41 percent of these). Rhone-Alpes had 11.2 percent of overnight stays (overseas – 34 percent) and Ile-de-France 10.4 percent of overnight stays (overseas – 78 percent). Between 22 French metropolitan regions, these three regions share more than 35 percent of the 1.5 billion of total overnight stays made in France (French – 64 percent, overseas – 36 percent) (French Ministry of Tourism 2006a). However, in terms of total tourism income (108.1 billion Euros including 92.4 billion Euros directly attributable to the regions), Ile-de-France is ranked first ahead of PACA (French Ministry of Tourism, 2006a) and is clearly dominant in terms of overseas visitors. This economic pulling power confirms the importance of the tourism industry and its policy priority in the City of Paris and Ile-de-France.

Summarising recent changes in French tourism during the last decade, we should pay close attention to the costs of the energy sector that have caused economic fluctuations. The French tourism industry was seriously affected by the rise in the price of oil just after recording a growth of 36.8 billion euros in 1997. In 2000, it lost 13 billion euros. In 2001 and 2002, the balance was slightly improved but deteriorated again in 2003 because of the general decline in French trade, the impacts of 9/11, and the high value of the euro (French Ministry of Tourism 2006b). Even though the economy of the tourism sector maintains its significant position in the overall French economy, it continues to display a slight decline (in 2005, 6.4 percent of GDP).

The Tourism Economy in the Paris Region

The tourism sector in the Paris region can obviously be distinguished from the rest of the French regions. It is ahead in terms of tourism income, on average, 237.20 euros per day were spent by each of the 26 million tourists who visited Paris in 2005. Tourism generates more than 500,000 direct jobs

(Chambre de Commerce et d'industrie de Paris 2007c). This area attracts the greatest number of foreign tourists to France.

Evidently the role of the City of Paris itself is crucial for these results. In business tourism (44 percent of hotel stays in Paris) as well as in tourism for leisure (56 percent) (Chambre de Commerce et d'industrie de Paris 2007c), Paris still holds the position of a world leader because of its museums, history, cultural attractions, events and festivals, hotel capacity (in 2004, 25 percent of French rooms and 31 percent of number of stays are concentrated in Paris region hotels (Institut d'Aménagement et d'Urbanisme de la Region d'Ile de France 2005)), and international airports. After the City of Paris, the Department of Seine-et-Marne accommodates about 15 percent of foreign visitors, in large part because of the attraction of the Disneyland Paris Resort. The proportion of short stays (fewer than three nights) is also a particular characteristic of the Paris Region. In fact, 70 percent of overnight stays are short stays in Ile-de-France against 56 percent in France (Institut d'Aménagement et d'Urbanisme de la Region d'Ile de France 2005).

Here we present two tables, Tables 3.2 and 3.3, showing the most visited tourist sites in Paris Region. In terms of the attractiveness of Paris region tourist sites, according to Table 3.3, the disparity between the Départements in Ile-de-France is an issue that deserves particular attention. We can see the massive concentration of visitors in the City of Paris. We count at least seven sites attracting more than two million visitors while in other Départements, only Disneyland and Versailles can attract as many visitors as Parisian historic sites. The Département of Hauts-de-Seine, the richest in France, attracts very few visitors, although some are attracted to the Grande Arche at La Défense, the business quarter. In Val-de-Marne and Val d'Oise, it seems that tourist attraction is very weak, even though these suburban areas are very close to the Paris core. However, as we shall see later, these Départements are carrying out a series of attempts to develop several local tourism sites.

Table 3.2 The most visited sites in Paris in 2005 (sites attracting more than 2,000,000 visitors)

Sites	Number of visitors	Evolution compared to 2004 (percentage)
Notre-Dame de Paris	13,000,000	+1.6
Sacré Cœur de Montmartre	8,000,000	0
Musée du Louvre	7,553,000	+14.4
Tour Eiffel	6,428,441	+3.2
Centre Pompidou	5,341,064	0.6
Cité des Sciences de la Villette	3,186,000	+14
Musée d'Orsay	2,918,225	+9.6

Sources: Chiffres clés de la région Ile-de-France edition 2007, Chambre Regional de Commerce et d'Industrie Paris Ile-de-France, 2007.

Table 3.3 The most visited sites outside Paris in 2005

Sites	Number of visitors	Evolution compared to 2004 (percentage)
Disneyland Resort	12,300,000	−0.8
Château de Fontainebleau	315,718	−0.7
Domaine de Vaux-le-Vicomte	237,762	+9.6
Domaine de Versailles	3,446,881	+4.4
Parc zoologique de Thoiry	384,000	+6.1
France Miniature	210,000	NC
Espace Rambouillet	105,844	+9.3
Verrerie d'art de Soisy-sur-Ecole	114,453	−10.9
Grande Arche de la Défense	198,048	+19.4
Musée de l'Air et de l'Espace	235,290	+19.9
Basilique Saint-Denis	141,354	−2.5
Stade de France	90,937	−9.5

Sources: Chiffres clés de la région Ile-de-France edition 2007, Chambre Regional de Commerce et d'Industrie Paris Ile-de-France, 2007.

The Evolution of Business Tourism

The number of business tourists is particularly important for the regional economy. Nationally, business tourism accounts for only 4 percent of visits, but in the Paris region about 13 percent of visitors make their trips for business purposes. The Paris region concentrates almost 80 percent of business travellers to France (in 2005) of which 60 percent are concentrated in the City of Paris (Chambre de Commerce et d'Industrie de Paris 2006). In 2007, for the twenty-seventh consecutive year, Paris was the top Congress City in the world according to the listing produced by the Union des Associations Internationales (UAI) (Chambre de Commerce et d'Industrie de Paris 2007c). All these figures make clear the important role of business tourism in the Paris regional economy, with the turnover of 4.5 billion euros in 2005.

However, competition is particularly fierce in this field. For both personal and business trips, global competition with serious rivals such as London, Barcelona, Milan, some German cities and more eastern European cities (Vienna or Prague) has become more and more intense. In addition to these European competitors, many cities in developing countries, in particular Asian cities including Dubai, Peking and Shanghai, threaten the existing European tourism market. The UN World Tourism Organization (UNWTO) forecasts that the Old Continent which attracts 55 percent of international tourism currently, should face up to a plausible decrease of its share to 46 percent in 2020 (Chambre de Commerce et d'Industrie de Paris 2007c).

Through this brief review on the specifics of tourism in the Paris region it is apparent that the region should prepare itself for future challenges, adapting its offer to the new requirements of the world tourism market such as improvement to the reception of visitors, modernisation and expansion of congress and exhibition capacity, use of new technologies, and the creation of new events. These conditions are acknowledged as important prerequisites to survival in the face of intensifying competition. In particular, the production of new attractions beyond the classical Parisian circuits will be one of the most important challenges for the Paris region whose general image is more or less fixed at its old historical sites (Notre Dame, Eiffel tower, Montmartre, Versailles, etc.). The development of new paths in the Paris region will be discussed later in the chapter. First we need to look at the complex interactions of the different scales of the government of tourism in Paris and Ile-de-France.

NATIONAL AND REGIONAL TOURISM AND URBAN STRATEGIES

In an atmosphere of growing competition and economic tensions, tourism development policies are being reorganised in order to improve and re-brand France's national and local image in regard to tourists and to improve the system of governance. The French governance system for tourism is composed of public authorities at different scales (from national to local) in collaboration with various private stakeholders (from civil associations to private companies). Sometimes criticised for the lack of communication between public and private actors, the aim of this system of governance has been to enhance internal coherence and to operate as a dynamic actor for more sustainable tourism development.

National policies

The role of the national government is fundamental in the enhancement of national infrastructure for tourism (transport, communication networks, education and careers etc.), to coordinate tourism strategies at a broader scale, and provide an essential part of subsidies for sub-national governments and other private stakeholders. The French Ministry of Tourism defines the principal challenges facing French tourism development as:

1. acceleration of the growth in world tourism and the opening of new destinations including China, South America, Central Asia etc.;
2. severe competition for the quality of tourism products with European rivals of France and aggressive enhancement of their national policies for tourism in recent years (e.g. Spanish Quality Plan, Brand Italia in Italy, etc., these are also very well-subsidised);
3. industrialisation of production and distribution networks due to globalising tourism activities;

4. reorganisation of working hours and new mobilities due to fast and low-cost transport;
5. rise of constraints for security due to 9/11 and recent London attacks;
6. ethical constraints that could have increasingly greater effects on the orientation of tourism development.

(French Ministry of Tourism 2006c, authors' translation)

In response to such perceived challenges the major orientations set out by the government in its 'Tourism programme for 2007' can be summarised as:

1. reinforcement of attractiveness to tourists through a 'Tourism Quality Plan' that aims mainly at the encouragement of tourism employment and the promotion of the image of France for the French people and foreigners;
2. promotion of social tourism policies in order to support improvements for particular publics (seniors, poor families, disabled persons, etc.);
3. pursuit of 'contract' between state and region (Contrat Plan Etat-Région) and creation of new local contracts for a better governance system;
4. modernisation of government support through various public organisations such as Groupement d'Internet Public Odit France, le Conseil National du Tourisme and its regional delegations.

(French Ministry of Tourism 2006c)

These principal orientations of national policies have been set out with the aim of more efficient promotion in the context of global competition, better contractual collaboration between local and national institutions for better public governance and a restructuring of national organisations. Increased competition has given the national scale a new concern with the tourism sector and new interest in how sub-national governments should respond.

General features of Paris Region tourism policies

The greatest issue for regional tourism is the relationship between the City of Paris and the other parts of the region. We have already briefly examined this issue through the statistical data reviewed in the previous section. The most discussed problem of tourism in the Paris region is that the rest of the region is much less well known as a tourist destination because of the predominance of Paris. In addition, the overwhelming clichés of classical and historical Paris remain firmly rooted in the imagination of all foreign and domestic tourists. This has inspired public authorities to identify the following challenges: 1) balancing Paris and the rest of Ile-de-France (lessening the 'predomination of Paris') to encourage enhanced tourist flows and establish a broader vision for city–suburb–region relationships, and 2) finding new paths off the beaten tracks in Paris and in other parts of the region through contemporary images (less 'past historical images') through the promotion of living and creative cultures (less 'frozen images') and through the presentation of a convivial and human background to the agglomeration.

One way of achieving such interdependent aims is through the development of a local niche tourism market seeking new alternative tourism destinations and the creation of new consumption spaces. Such attempts may stimulate more sustainable and participative tourism and encourage revisits from tourists. To take up these challenges, a series of pilot actions (e.g. enhancement of tourism infrastructure and quality of reception, installation of tourism related institutions, etc.) are being undertaken by the regional government.

Priorities for Paris Region tourism development

According to the CESR (Conseil Economique et Social Région Ile-de-France) report on the tourism and leisure policies of Ile-de-France for 2000 to 2010 (French Ministry of Tourism 2006c) the most important issue is obviously addressing the image of Paris and the rest of the region and the attempt to present them as a single entity. For example, a set of strategies have been developed for cultural tourism across several suburban sites such as Versailles, Fontainebleau, Rambouillet and Vaux-le-Vicomte in order to rebalance tourism given the crowded sites concentrated in Paris 'intramuros' (see Tables 3.2 and 3.3).

The Paris region plays the role of a leader in the global tourism market due to the strength of foreign visitor numbers. In 2005, there were more than 150 million overnight stays (78 percent foreigners) which accounted for 10.4 percent of the total overnight stays in France. By contrast, the weaker point of the region is domestic tourism; the market share of the Paris region has fallen from 9.6 percent of about 400 million overnight stays made by French residents in 1990 (Baudoin 2000), to 4.9 percent of 826.6 million overnight stays in 2006 (French Ministry of Tourism 2007). We can also see a small decrease in the average length of stay and the frequency of visits of French tourists. In addition, we can observe an unstable fluctuation in business tourism due to the rising competition with other French and European cities such as Lille, Lyon, London, Brussels, etc., although for the moment Paris still preserves its advantage.

The tourism market in the region is subject to various external economic circumstances such as volume of visits closely related to events (fairs, exhibitions, congress, etc.). Concerning the strategies for niche marketing, the mainstream policy of the region is to encourage industrial and scientific markets in order to attract more business travellers and young people (notably students and young professionals). Consequently, the revival of the 'festive' attractiveness of Paris through the creation of events (festivals, sports events) and multiple attraction spaces for culture and entertainment has also been given greater priority. New initiatives such as *Paris Plage*, *Paris Respire*, and *La Nuit Blanche*, along with new cultural festivals in Paris, support these new priorities for tourism promotion. In this context, how can Paris attract tourists to visit and revisit the region? It achieves this through more targeted communication, a more diversified tourism offer, more management in order to raise the ratio of quality:price:satisfaction of tourists and more organised transportation and public access.

The regional government imagines a series of prime regional tourism clusters within Ile-de-France (Institut d'Aménagement et d'Urbanisme de la Region d'Ile de France 2006). These are sites considered to have new potential for tourism, and are managed by the regional and departmental governments in collaboration with national government to respond to new tourist demands, for example health, discovery, adventure and culture. The clusters are outlined in two planning documents, the Regional Scheme for Tourism and Leisure and the Plan Contract State-Region (2000–2006). At present, we can count eight such clusters in the region.

Here, we will focus on two clusters, that of north-east Paris and that of the City of Paris itself. They will help illustrate the recent evolution of Parisian tourism stressing the trend towards alternative and sustainable tourism, the collaboration between Paris and suburban areas and the desire from tourists for new consumption spaces. We examine Belleville and Bercy, included in 'the Parisian Cluster' in our case studies (see Figure 3.1).

The Parisian Cluster

The main themes of the Parisian Cluster are to stimulate participative tourism and synergies between neighbourhoods, encouraging the encounter between Parisians, Franciliens (this term refers to the population of Ile-de-France excluding Parisians) and external visitors through the development of places of participation (locals and visitors), home stays, urban discovery promenades, etc.

The important sites in the cluster are 1) North-west area – Montmartre, the flea markets, Belleville and La Villette; 2) South-west area – Porte de Versailles area, André Citroën Park, the Bois de Boulogne; 3) South-east – Bercy and the National Library of France, the Bois de Vincennes and its Floral Park.

North-east Parisian cluster

The north-east Parisian cluster lies in the department of Seine-Saint-Denis and is captured by its marketing slogans: '*Two steps from the heart of Paris, a territory of international events, a traditional suburb and laboratory of creativity*', and '*Five different sites on one axis to redesign the Paris of events*' (authors' translation). The cluster is formed around large cultural, tourist and leisure facilities (Stade de France, the Basilica of Saint-Denis, the Musée de l'Air et de l'Espace, la Villette, the Parcs des Expositions du Bourget et de Paris-Nord Villepinte, and les Puces de Saint-Ouen) already established as attractions in the area, shaping a clearly visible image for local stakeholders and international visitors. This cluster is seen to have great tourist potential, stimulated by the on-going urban regeneration projects of Paris northeast around the north-east sector of the peripheral motorway. The benefits to the local image of the designation of the cluster are expected to include local urban and economic regeneration and a strengthening of the local population's sense of belonging inducing a new positive perception of the value of their heritage.

In this section, we have observed briefly the Paris region's evolving tourism policies. The new regional plan for Ile-de-France, the 'Schéma directeur de la région Ile-de-France' (www.sdrif.com), summarises the orientation of key policies that we have examined: 'It is therefore necessary to combine the preservation of the existing attractiveness of the Paris Region and the diversification of new tourist flows, towards alternative practices and places of cultures' (Chambre de Commerce et d'Industrie de Paris 2007a (authors' translation)) . However, whilst there appears some consensus about the direction of policy it is not an easy task to clearly discern who bears the main responsibility for tourism policy. For the City of Paris, which also possesses the status of 'Département', the City government is the central actor in relation to tourism policies and other relevant policies. For other communes in the region, competence in tourism policy is shared by different territorial scales (commune, département and region) and always supported, and at the same time controlled, by the national government. In addition, there are a great number of major private stakeholders including the tourism industry and local civil associations. In particular, subsequent to the rise of participative and alternative tourism, and the emergence of a more 'democratic' planning process, the local population's voice is becoming stronger and stronger in various ways, supporting or disagreeing to the development of tourism in their territory. Certainly, the key to the success of both urban and tourism policies would seem to lie in the collaboration and synergy between different stakeholders sharing different visions and methods in the style of the French 'urban project'.

THE URBAN PROJECT, TERRITORIAL COMPETITIVENESS AND TOURISM DEVELOPMENT

To understand more about how these new approaches to tourism may develop we turn to a broader discussion about approaches to urban development.

How can tourism policies be articulated through the complex and delicate relationships between 'urban project', 'competitiveness' and 'attractiveness'?

Strategies around urban projects have guided French urban policies since the end of the 1980s (Ingallina 2001). The use of the term 'project' signifies a change in the mentality of French planning legislation (e.g. the creation of SRU law about urban 'solidarity' of December 2000) compared to the previous notion of a fixed, rigid, essentially prescriptive 'urban plan'. Decision-makers prefer to employ the term 'project' instead of 'urban plan' because of its flexibility. A project is a process of creating 'something': an image, an urban space, a strategy or a partnership between different actors, etc. As an open process, it allows different kinds of negotiations with the public.

The project approach (Ingallina 2001) in the French planning system is based on the 'articulation' between formal and informal planning tools, involving many stakeholders, many competences, partnerships and other forms of negotiation. It is also based on the idea that planning is an open process proposing a global and inclusive vision of territorial development in the turbulence of urban growth (Bouinot 2002). It is flexible, considers multiple temporalities (Roncayolo 2003) and multiple and changing partnerships (with various forms of negotiations). It also follows a qualitative approach – bringing prospective scenarios and allowing a shared representation of a territory combining the past and the future; its current identity and what it will become (Ingallina 2004). To understand territorial strategies in France, it is necessary to comprehend this complex approach, mixing various methodologies and competences, at different scales and temporalities and involving different actors.

Territory and attractiveness

Overall, recent project strategies demonstrate two principal objectives as described below.

1) Increasing the performance of a territory in order to improve its competitiveness. Competitiveness is not only related to the performance of companies but it involves also the French concept of 'attractivité' (Ingallina and Park 2005, and Ingallina and Alcaud 2008) that means the capacity of a territory to be an attractive point in 'global flows'. Territory is not only conceived geographically but is also viewed as a symbolic referent – an image with many dimensions emerging through people's shared and complex visions.

2) Encouraging social and environmental solidarity to improve citizens' quality of life. For example, urban policies more and more take account of the crucial role of public spaces in a city and, as well as their urban design quality, they are conceived as a part of the marketing to improve the image of places. Public space is seen to play a particular role in forging solidarity through incorporating different functions and allowing social-cultural exchanges. There is some affinity with the EU concept of 'territorial cohesion' that means, here, not only social equality (all forms of social and cultural solidarity) but also an entity that a territory can offer to its inhabitants, in

French, 'offre territoriale' (Hatem 2004). This includes offering quality of public spaces, mixing global and local values, providing new facilities and services in order to attract global flows of visitors to stimulate consumption (Park 2005a) with economic effects (e.g. job creation). In this sense, developing a competitive attitude for a territory by means of a 'territorial offer' is seen as compatible with social solidarity and principles of environmental sustainability, if it can make an adequate balance between them. City image, important in attracting visitors, is also a vital part of the conception of an integrated urban project.

PLANNING ISSUES IN THE PARIS REGION

This perspective on planning French cities has a particular character in Paris and Ile-de-France. The principal planning tools, the SCOT (Schéma de Cohérence Territoriale at region scale), the PLU (Plan Local d'Urbanisme at local scale) and the PADD (Projet d'Aménagement et de Développement Durable, connected to both of them), assist the region to enhance competitiveness, not only in relation to investors but also in relation to visitors, artists, students and researchers. For that, local leaders often adopt policies based on design (*projet*) in order to make the territory more 'attractive' and then more 'competitive'. In the Paris region, in particular the inner *zone dense* (with about 5 million of the inhabitants of the region as a whole) the aim is to increase economic competitiveness and job creation by means of a system of 'clusters of competitiveness' combining research and production. Tourism is also considered as a major area of action to develop the Parisian economy – through enhancing the environment, developing cultures and employment (PLU *Diagnostic* 2004). As a result, meetings, mega-events or media-events (such as the Olympic Games or the Rugby World Cup) become more and more an *objet du désir* for all municipalities, small or great, principally in the first ring of Parisian *banlieue* (roughly the four inner departments shown on the map). In the previous sections we have looked at this tourism–urban planning relationship and some examples of 'pôles' (or 'clusters') for tourism development in the Paris region. In addition to tourism objectives the PLU of Paris also emphasises the concept of 'territorial offer' as part of its effort to reduce territorial inequalities providing urban landscapes of better quality, public spaces adapted to specific quarters (for example the most disadvantaged), and provision of necessary services and infrastructure.

As we have argued, the different scales of government and intra-regional differences create some specific issues for Ile-de-France region. The City of Paris, for example, has an exceptional economic and cultural role and has not yet begun looking for a real partnership with neighbouring municipalities that have tried, separately, to become more attractive and competitive by taking advantage from their regional location. Some municipalities located in the first ring of the Parisian *banlieue*, such as the municipalities of Val-de-Marne, have an individual vision of territory. Establishing administrative partnerships near any capital is always very difficult.

The City of Paris has also realised a PADD at urban scale, but it has provided some *Projets de Territoires* in partnership with cities of the first *banlieue* ring in a so-called 'GPRU' (Grand Projet de Renouvellement Urbain) document, with particular regard to the territories located near the *Portes de Paris*. In addition, the PLU of Paris attempts to strengthen the cooperation between Paris and its agglomeration, in a large range of actions including the GPRU's social housing and public spaces, and the PDU (Plan de Déplacement Urbains) plan of urban mobility of the Ile-de-France region. Outside the City, the development of the north-eastern suburbs includes cooperative projects around *Plaine Saint-Denis* (immediately north of the City) and a broader negotiation with the municipalities of *Plaine de France* (an area extending from the City to Roissy airport) and wider environmental initiatives such as the PRQA (Plan Régional pour la Qualité de l'Air), etc. Such initiatives reflect some development over the past few years on the central question for planning in the Ile-de-France region – how to find the right place for Paris, in the centre of this agglomeration, in order to share equitably all benefits that its territory is able to produce. There is much debate about getting together actors in the Parisian agglomeration to enhance economic efficacy and to encourage its urban and social dynamism. The political and institutional challenge is marked in the urban landscape. The connections of the River Seine into the *banlieue* create many particular landscapes and opportunities for cooperation. Where there is a physical barrier to cooperation, for example the *boulevard périphérique*, some effort has gone into studies of crossing points in a series of urban projects. Furthermore, the 'Contrat de Plan Etat Région' (2000 to 2006) established ten *pôles de croissance*, three of which are connected to Paris: the Plaine Saint-Denis, in the north, the Seine Amont, in the east and the Val-de-Seine, in the south-west. In particular, the départements of Seine-Saint-Denis and Val-de-Marne have launched a programme on 'industrial and cultural tourism' (see, for example, the MAC VAL – Museum of Contemporary Art, located in the City of Vitry-sur Seine in the Department 94, Val-de-Marne) specifically addressed to local inhabitants (Conroux 2007).

There is a structural issue of developing projects across administrative boundaries and an issue of integrating policy objectives. Tourism is increasingly seen alongside economic and social planning. In addition, one of the main priorities of tourism development policies in the Paris region is to enhance the physical quality of many urban spaces. The *embellissement* of Paris is the first objective of the PLU, and illustrated by the PADD. The City of Paris continues to develop tourism initiatives in different ways: emphasising urban classical heritage and the renewal of the landscape of the banks of the Seine, but also promoting the enhancement of other urban landscapes such as Belleville, and creating new urban landscapes of consumption like Bercy Village. The image of places is also enhanced through many temporary events such as *Paris Plage* or *Paris Respire*, encouraging walking and the use of bicycles instead of cars. The 'festivalisation' of the City familiar in North America has also become one of the efficient promotion strategies of Paris.

In Ile-de-France, and particularly in the urban core, we can see evidence of the urban project approach to urban quality and renewing City image and a new interest in connecting policies across the administrative and physical barriers. Tourism policy is readily incorporated into this overall approach to urban planning. However, there are also some fundamental economic challenges facing the region that impact on these new approaches to urban policy and at the core there is a perceived problem, not about the flows of visitors, but about maintaining the region's relative wealth and middle-class.

The French economist, Laurent Davezies raised this question in his study for CDC (Caisse des Dépôts et Consignations) in 2007 with a provocative title: *Croissance sans Développement en Ile de France* (Davezies 2008). Davezies was concerned about the risk of a process of under-consumption in Ile-de-France. Ile-de-France is a paradox; on the one hand the region is one of the most important places of production (just after the cities of Tokyo, L.A. and N.Y.) with a GDP in 2003 of 513 billion dollars for 11 million of inhabitants (that is 40 percent of Chinese GDP!). On the other hand, the incomes of 'Franciliens' have been decreasing since the 1980s. In 1976, the region produced 27 percent of national GDP and its residents shared 25 percent of national household incomes. Currently, the region produces 29 percent of national GDP but with only 22 percent of household incomes (Davezies 2008). There has also been considerable job losses in Paris. In the last ten years, Paris have lost 126,000 paid workers (93,000 workers in the industrial sector and 33,000 in the building sector, according to INSEE). As well as young workers, the region also loses retired people who want to live in a more pleasant environment than Ile-de-France.

There is also a gap between the income of the residents of Ile-de-France and consumption because workers in Ile-de-France are increasingly tending to spend their money in other regions. So we can say that even though Ile-de-France still attracts a lot of foreign tourists, and they spend a great amount of money, this region loses part of its wealth because of a loss of expenditure from local people. Thus, a crucial question for planning policy becomes how to balance the policies and projects designed to attract more tourists with the necessity to maintain local people.

In this section we have summarised the urban planning process in France and examined the case of the Paris region that introducing some problematic issues concerning the future of tourism. Nonetheless, local authorities in the region consider that culture and tourism are the best assets for local development in a similar way to many other cities and regions all over the world. Such strategies are often considered highly efficient for the improvement of urban attractiveness and are usually based on a concept of urban consumption that includes culture, leisure, education and tourism. We can say that the emergence of cultural and tourist features in city planning is already widespread, even if its application in real and concrete projects is rather delicate and complex. Indeed, several experiments show that to improve the tourism aspect of city development it is not enough to conceive ambitious tourism promotion strategies. It is also necessary to simultaneously develop strategies that are based on a comprehensive territorial project with partnerships between various actors in this process. For this reason, it is important to understand the general background of the urban planning process of the Paris region. From this position, we can go on to analyse how the Paris region's tourism policies are developing and how they are articulated within its urban projects.

THE TOURIST CITY – TRADITIONAL ATTRACTIONS AND NEW PLACES

In this section we will discuss new developments in the context of the overall tourism offer in the Paris 'cluster'. For tourism planners, the charm of Paris should not be confined to its famous tourist circuit, 'Tour Eiffel, Louvre, Printemps, Galerie Lafayette'. Instead, it can be extended to undiscovered Parisian quarters and neighbouring communes providing original paths to visitors and allow them to discover the other sides to the development of the City. According to some forecasts, by 2020 Paris will receive twice as many tourists as it does today. These 50 to 60 million tourists will have an impact. There will be visitors from developing countries (Brazil, Russia, Asia . . .) most of whom will be visiting Paris for the first time, accompanied by a tour operator. We would expect them to make a tour of the classical tourist circuit. On the other hand, we should think about the tourists who have already visited Paris, appreciated their stay, and decided to return. According to Paul Roll, director of the Tourism Office: 'They are in search of a different Paris' (Chambre de Commerce et d'Industrie de Paris 2007a). On return visits, tourists become more independent and want to leave the beaten track in search of 'The Paris of a Parisian'. For some, this search for new experiences motivates them to meet local inhabitants and to share moments of local life (Sallet-Lavorel 2003). The rising costs and the worsening quality of the welcome at traditional tourist sites play an important role in this behaviour. If the managers of these sites hope to maintain their attractiveness they need to change the perceptions of tourists and improve the quality of reception and service. While traditional locations are improving their offer, we expect tourism demand to mean more emphasis on new areas.

New attractions in Paris and its outskirts

New and alternative forms of tourism are a rising trend of Parisian tourism as visitors search for a 'real' Paris. In many cases the most important point is to preserve and redevelop the unknown past of Paris. Historical images of Paris are regarded as an inestimable heritage despite criticism concerning Parisian clichés; how can Paris create an adequate mix of memory and innovation?

In some new places on the outskirts of the City of Paris tourism development can encourage the development of neighbouring communes – such as Saint Ouen, Ivry sur Seine, Créteil, and Charenton. In our case study of Bercy in the 12th arrondissement, new tourism may stimulate further economic development in the 13th arrondissement and nearby suburban areas such as Charenton and Ivry sur Seine in the département of Val-de-Marne, currently one of the least popular tourist destinations in the Region.

All the outer areas of Paris have assets to exploit. The neighbouring départements of Seine-Saint-Denis and Val-de-Marne have initiated development of their industrial heritage. The departmental committee of tourism in Val-de-Marne, in collaboration with the Chamber of Commerce and Industry of Paris and the agglomeration, invites tourists to come and visit various industrial sites with a guided visit programme entitled, 'Travels to the heart of industry' (Chambre de Commerce et d'Industrie de Paris 2007b). This new industrial tourism goes hand in hand with a new partnership approach to tourism development.

This trend towards alternative tourist circuits is furthermore stimulated by a series of urban regeneration projects for decaying quarters. In eastern Paris, for example, visitors can discover countless memories of arts, crafts, industries and their workers through its built environment and diverse population and activities. We look next at the cases of Belleville and Viaduc des Arts to observe briefly how the renewal or rediscovery of vestiges of old Paris can stimulate new forms of Parisian tourism.

Belleville: an old quarter for new participative and alternative tourism

Belleville is an old working-class quarter in eastern Paris. Traditionally a quarter of factories and craft workers, this area also preserves charming memories of popular musicians from the beginning and middle of the twentieth century (Edith Piaf, Maurice Chevalier, etc.). Unknown as a tourist path in Paris, because the area provides neither great monuments nor fine shopping streets, Belleville has recently become a new example of participative and alternative tourism in Paris. The quarter is promoted through 'Urban Discovery Walks' and Internet sites where Belleville tourist information uses the slogans: 'Discover the vibrant quarter of the outskirts of Paris . . .', 'Meet the locals, eager to share the secrets of their neighbourhoods . . .' and 'Feel the pulse of the City as you explore urban streets and hidden courtyards . . .'. Most well-known is a site created by a civil association, 'Ça se visite' (http://www.ca-se-visite.fr/) that organises regular visits to the Belleville quarter following a series of cultural themes. In addition, the City Council of Paris supports a particular project of the promotion of Belleville. The project involved creating an Internet site named 'Web d'en Haut', in order to preserve the memory of Belleville, to provide local news and events and to encourage exchanges of services between inhabitants and community associations. Tourism promotion is principally based on these convivial and popular aspects of the quarter. In addition, by proposing meetings with local artists, craftsmen and shopkeepers, the quarter promotes its assets as a way of offering a different experience that allows the discovery of the charms of a genuine and contemporary Paris. Tourists can encounter the distinctive housing architecture of the nineteenth century with small gardens and courtyards, the ateliers of craftsman and artists who are now becoming well known figures of the quarter. Another important feature of Belleville is that the quarter is filled with ethnic merchants, in particular having a substantial number of Chinese restaurants and groceries. Ethnic diversity is a very important component of the general landscape of the quarter, often called 'Babelville', but as with the exploitation of diversity in other cities, critics ask if they are encouraging diversity (actually Belleville promotes itself as the most cosmopolitan quarter of Paris) or weakening the authentic aspect of the quarter? That said, the promotion of Belleville is more the result of civil efforts to enhance the quarter as a convivial place for local inhabitants and other Parisians rather than the result of structured tourism policies.

Local government has just started to pay attention to this quarter and its new tourism potential. However, the quarter of Belleville is located between 11th, 19th and 20th arrondissements and this makes it difficult to establish an efficient public governance system. The quarter has been principally promoted by a 'spontaneous' civil participation. Up to the present, we can merely perceive some simple public sector urban interventions for cleaning, lighting, signboards and urban furniture etc. in the quarter signifying the public authorities are not yet deeply involved to the promotion of the quarter.

Viaduc des Arts and Promenade Plantée

This is acknowledged as one of the most interesting urban regeneration projects in eastern Paris. Starting from Bastille, and crossing the 12th arrondissement, this long urban walkway is accompanied by a very particular style of urban design. The Viaduc was an abandoned and decaying nineteenth century railroad viaduct that was planned to be demolished. Instead, it has been converted to a sophisticated shopping and urban park. This unique location is home to more than forty-five craft shops in the arches that support the renovated railway viaduct and the architects Philippe Mathieu and Jacques Vergely have designed a kind of aerial green parkway, the Promenade Plantée, on top of the arches. The new shops recall the history of the area that was once the home of craft and antiques shops. In the 1990s, new craft shops and bars were introduced. Under every arch of this viaduct lies a different aspect of art and consumption that can attract both locals and tourists. The design exploits heritage creating a contrast between Parisian architectural history with brick and arches, and the contemporary design of shops under the arches. The character of the older working-class neighbourhoods of Paris that is being lost in other places, is captured by the presence of the Viaduc des Arts, continuing its new life beyond its first usage despite some occasional criticism concerning discontinuity of the fine and sophisticated style of its shops.

Contrasting with Belleville, that is spontaneously promoted by civil initiatives, the Viaduc shows a strong hand of the City of Paris in regenerating eastern Paris to create a new consumption space. In the case of Bercy, we can analyse more concretely the ambitions of the City of Paris for its eastern area through urban regeneration projects.

Bercy – urban project and tourism quarter

In Belleville and the Viaduc there are questions about how Paris can reconcile the demands of tourist sites and the authenticity of Parisian quarters. Here, we will focus on the role of new consumption spaces in the transformation of the city image. Urban policy makers have become more and more aware of the role of consumers (including both inhabitants and tourists) as a new economic resource and source of local income. Of course, cultural and tourism consumption promotion strategies cannot be considered only as economic development strategies. Cultural and tourist resources could also generate social benefits, for example, cultural and educational services for the local population or support for a new creative industries. We can trace these attitudes through the case of Bercy Village where the urban, social and tourism impacts of a new consumption space have in turn had an impact on the changing image of the City of Paris.

Located in the 12th arrondissement, Bercy Village is a part of the urban operation ZAC de Paris Bercy which was outlined in 'Le Plan Programme de l'Est de Paris', a comprehensive renewal and regeneration plan for the whole eastern part of Paris developed by the City in the 1980s. It would be wrong to see the Bercy as an isolated commercial project as the long term planning of the ZAC de Paris Bercy aimed to produce projects that would create an authentic 'piece of the City' with the expectation of inducing urban revitalisation benefits for the whole of the east of the City. The new Bercy quarter has come to represent the revival of the eastern part of Paris that was vastly underdeveloped compared to the west. Bercy's traditional wine trade and its vast riverside warehousing was in decline by the 1960s, and a comprehensive scheme for 'Secteur Seine Sud-Est' was approved in 1973, this was the first attempt at a strategic perspective for urban renewal. It involved the creation of a new tertiary employment to this large area (280 ha), emphasising housing development. At that time, the site was almost entirely vacant and offered exceptional opportunities for redevelopment. (APUR 1974). In 1983, 'Le plan programme de l'Est de Paris', one of the flagships supported by Mayor Chirac proposed a new image of the capital. The plan officially fixed the Bercy operation in the Zone d'Aménagement Concerté (ZAC) de Paris Bercy. This was a multi-use programme including large scale cultural facilities (a huge cinema multiplex, the Palais Omnisport de Bercy, the National Cinematographic Centre) making one of the most high profile developments in eastern Paris. Bercy has contributed to new images of Paris as a whole through the marketing of itself as a young and dynamic residential and commercial quarter to Parisians who seek new places to live, visit and work. The new business quarter within the ZAC Bercy has also attracted a large number of companies in the communication and new technology sectors.

Bercy Village is often cited in the French media as the first Urban Entertainment Centre in France with its thirty establishments (multiplex, shops, restaurants, fitness club, etc.). It provides citizens with many new cultural and leisure activities, restaurants and a sophisticated shopping gallery. Its promoters claim inspiration from the experiences of London's Covent Garden and South Street Seaport in New York. According to the site's management company, ALTAREA, Bercy projects this new approach to consumption spaces. It draws on the atmosphere and traditions of old wine warehouse buildings and the cobblestoned main street claims some authentic architectural appeal.

Since its opening in 2001, Bercy Village has been a commercial success with 4.5 million visitors and an initial turnover of 68.6 million euros in its first year of opening (Allaman 2002). According to a study by CSA in 2002 entitled, 'The study of Parisian perceptions regarding life in their neighbourhood', the site attracted 6 million visitors in 2002 (35 percent more than the first year) and 96 percent of local inhabitants in the 12th arrondissement of Paris (where Bercy Village is located) expressed satisfaction with the new project. In the wider region, by 2003, 44 percent of the population knew about Bercy Village.

The customer base is drawn from the immediate residential and business neighbourhoods, and visitors from further afield are drawn to the cinema multiplex and the Palais Omnisports. Some recent customer surveys of Bercy Village (ALATREA 2003) identify the origins of the 5 million visitors as including workers from the ZAC Paris Rive Gauche that is located on the opposite bank of the Seine, visitors of the National Library of France who represent 1.5 million people, from neighbouring arrondissements and Franciliens from nearby suburban cities such as, Saint-Mandé, Charenton, Ivry, Le Kremlin-Bicêtre, etc. In addition, Bercy Village aims to attract about 5 percent of the total number of tourists in Paris. The hotels in ZAC Paris Bercy already draw more than 550,000 people to this part of the City and the presence of tourists becomes more and more important in Bercy Village (see Figure 3.2).

At present, Bercy Village is considered as a new place of conviviality for Parisians who represent 50 percent of the visitors, in particular, residents from the 12th, 13th and 14th arrondissements that are often considered 'boring'

Figure 3.2 Paris: Bercy Village
Source: Authors' own

in terms of leisure and entertainment places. A further 37 percent come from the Paris region, 11 percent of visitors are domestic tourists from other regions, and 2 percent are foreign tourists. During holiday seasons there is a higher volume of tourists. The share of foreign tourists is steadily increasing through the active promotional activities of ALTAREA in various travellers' guides and word of mouth effects. The average age of visitors is 37 years and they tend to have strong purchasing power (more than a third are executives and freelance workers) spending in general 60 euros a visit.

The area's attraction draws on the renovation of the City's industrial heritage, the old wine warehouses creating the nostalgic atmosphere of a small rural village in the nineteenth century. Bercy also has geographical advantages. The quarters of the 12th arrondissement of Paris are closer to the core quarters of Paris (Bastille, Notre Dame, etc.) than the majority of other regenerated eastern quarters. Bercy's attraction also comes from the mix of uses offering to satisfy the requirements of a range of consumers. The desire for festive and extraordinary scenery has also become important in daily life (Park 2005a). Such desires for 'virtual' realities (i.e. aestheticised and derealised) are interpreted by Gottdiener (Gottdiener 2001) as a kind of confusion deliberately created by the combination of reality and artificiality. Such demand generates the phenomenon of 'the festival setting' (Short 1996), that attempts to create a dazzling festival mood in a town. Unlike our other examples in the City of Paris, Bercy Village reflects a conscious desire to reproduce an urban festival setting.

CONCLUSIONS

To understand the case of Bercy we have to understand the role of culture and tourist consumption in improving urban attractiveness. Bercy is a symbol of the transformation of an outmoded industrial past, emphasising a diversified economy of culture, leisure, and new technology. The image of Bercy Village as a space for new urban consumers undeniably contributes to the embellishment of eastern Paris and the improvement of local image. This place is very visible, easily accessible and well communicated by mass media. This visibility creates an excellent asset for the urban marketing that contributes to the success of the urban project ZAC de Paris Bercy and tourism development strategies for the whole of eastern Paris. Originally conceived as a convivial local place, Bercy was not a tourist destination. In the City of Paris, filled with consumption places for tourists, above all, Bercy Village was seen as a new place for conviviality and distraction for Parisians. However, as we have seen, tourists have discovered the new eastern side of Paris. Many local cultural events (often for children) are held in Bercy Village associated with the Park of Bercy and those events, mostly intended for inhabitants of the 12th arrondissement are quite welcomed by this population as well as tourists. In addition, the 12th district has not been richly equipped with cultural facilities and events compared with other central districts in Paris because it is mostly residential. Moreover, the success of Bercy Village encouraged other programmes of multiplex-leisure complexes in nearby municipalities ('MK2' in the 13th arrondissement, 'Pathe' in Ivry sur Seine), multiplying new tourist paths as well.

We have emphasised the role of new consumption spaces in urban strategies for local development including tourism. The question of territorial identity and image is very crucial for all cities faced with global competition. Branding strategies have become commonplace as local policy makers realise the importance of local image to attract people, investment and economic activities. A competitive City owes its success to the improvement of attractiveness, through the creation of positive local images (Park 2005a). Since the 1990s, consumption oriented strategies have come into the mainstream of contemporary urban planning. Local benefits can arise as a result of such strategies, and the profitability of Bercy Village feeds into local community housing projects. Nevertheless, it should be remembered that these new spaces of consumption can also generate social conflicts through the dominance of expensive shops and restaurants and mechanisms of social control deployed to police such new places (Ingallina and Park 2005). A place like Bercy Village is generally identified as a public space through the flows and exchanges allowed in its area. However, we can clearly see the rising tendency to control public space in order to improve its security and aesthetic aspect.

The new tourism projects in eastern Paris and in some of the industrial suburbs show new attitudes to tourism planning and a city working with the pressures of demand, both on its traditional attractions and for new urban experiences. The tourism project needs to be understood in a comprehensive approach to city planning. The urban project has economic and social objectives and displays a new concern for image and marketing. The types of urban spaces preferred by consumers in eastern Paris reflect an image of a contemporary city mixing new design with heritage, new experiences with urban memory and an alternative to the traditional tourist circuit and clichés so important to the City's brand for so long. These new sensibilities reflect concern about tourism numbers and competition in a global market. The opening up of eastern and old industrial Paris is necessary if the City is to continue to be competitive.

4 London

Tourism moving east?

Robert Maitland and Peter Newman

As a centre of trade, source of finance for global markets, a city of government and sometime imperial capital, London attracts visitors drawn to the City for a range of reasons. The stock of historic buildings and 'zones of prestige' where, 'culturally impressive activities are produced, displayed and consumed' (Maguire 2005, p. 16) that are concentrated in the centre are the main attraction and reproduced in iconic images of the capital. Other activities, for example national and international sporting events draw visitors, and the underlying consideration of the bid for the 2012 summer Olympics was the economic value this presented in terms of increased tourism and boosted city image. The multiple attractions of the City mean that we need to carefully differentiate city visitors in order to understand how London is developing as a world tourism city. The multiple functions of the City – global financial centre, home of national cultural institutions and seat of government – mean that we need to examine the relationships between tourism and other forces impacting on its social and economic life. In the first part of this chapter, we review contemporary trends in tourism in London and focus on the changing flows of local, domestic and international visitors. In the second part, we review how tourism is understood in public policy and focus in particular on the objective of drawing visitors away from congested central areas. We then examine two 'new tourism areas' close to the centre where distinctive groups of visitors have contributed to wider processes of change and the transformation of the local landscape. We question the potential lessons from these experiences and the future of this type of urban transformation. The final section of the chapter reflects on potential future directions of change in London tourism.

London is well understood as a dynamic financial centre and capital city with well known visitor attractions. In recent years, it has become common to understand the City through its diversity. We should see the mobility of visitors alongside the flows of migrants in and out of the City and in the context of the social and economic transformation that marks these global trends. More than 300 languages can be heard on London's streets and some fifty ethnic groups live together – in harmony according to many (for instance Johnstone and Masters 2006) or barely tolerating each other according to others (for instance Benedictus 2005). Also helped by the global use of the English language, London has become a strong magnet for immigrants from all over the world, to the point that its ethnic minority population (all groups other than white) grew by 54 percent between 1991 and 2001 (ONS 2007a). As Benedictus, following his intercultural and culinary journey around London celebrates, 'every big city in the world has a Chinatown, but in London, one can dine on food from more than 70 different countries' (Benedictus 2005).

However, despite London's dynamism and the fact that earnings are on average over 20 percent higher than in the rest of the UK (ONS 2007b), the elevated cost of living, low perceived safety and notorious unemployment problems (7.1 percent of unemployment, the highest rate in the UK) do not lead to London being the first choice for living among British citizens (ONS 2007b). Notably, according to the 2001 census: 'London is a magnet for young people from all parts of the UK and the rest of the World but is generally less attractive to families and the elderly' (ONS 2007a, p. 10); in 2005, for instance, the capital region saw the largest net migration loss of the UK, with 81,500 Londoners moving to other parts of the country (ONS 2007a). For forty-nine years, between 1939 and 1988, the population of London declined following the peak of 8.6 million in 1939 and reaching its lowest point in 1988 with a population of 6.7 million (ONS 2007b). In the last two decades, however, thanks largely to the strong international immigration, there has been an increase of about 49,000 per year (ONS 2007a), that led to a population growth of 10 percent between 1991 and 2005 (a higher rate than the UK average) (ONS 2007a). These processes of transformation of the City as a whole give a context for trends in the flows and characteristics of visitors to the City.

TOURISM TRENDS

As well as being a focus for immigrants on a global level, London constitutes a major attractor for both national and international visitors, representing a fundamental source of tourism for the United Kingdom thanks to its role as a main gateway to Britain.

According to the London Development Agency (2006), three out of four overseas UK visitors pass through London, and 37 percent of overseas visitors to London visit other parts of the UK during their stay in the capital. On the other hand, the opposite is also true, visits to London depend to a certain extent on the success of other UK destinations, for instance Edinburgh or Stratford-Upon-Avon, with visitors to these destinations stopping in London on their way.

With over 26 million overnight visits each year (VisitLondon 2007), tourism represents today the second (behind financial services) most important sector in London's economy (12 percent of GDP) and, according to the London

Development Agency's estimates, it will become the first sector in terms of value by 2012. Tourism induced jobs are not easy to estimate since spending is spread across a wide range of activities (London Tourist Board and Convention Bureau 1997). However, according to London's tourism marketing agency (VisitLondon 2005a) 320,000 people are directly employed in this sector (13 percent of the total workforce) and a 1 percent sustained increase in overseas visits to London corresponds to 1.28 percent increase in London jobs (0.34 percent for domestic visits).

The total spending of overnight visitors to London in the last six years has been essentially stable at around £9bn per year (VisitLondon 2007), with overseas visitors bringing in about three times the amount of domestic visitors. Overseas visitors buy 30 percent of theatre tickets and account for half of all visits to London attractions. To these spending figures we can add a further £4bn brought about by the 150 million day trippers who visit London every year (VisitLondon 2005a).

Visitor numbers for the past six years (VisitLondon 2007) seem to confirm a trend of an increasing attractiveness of the capital outside the country while its appeal appears to be constantly decreasing among British citizens. Whereas seven years ago domestic trips to London exceeded overseas visits of about 5 million, today the opposite is true, with the switch occurring in 2004. Overseas visits increased from 13.4m in 1996 (London Tourist Board and Convention Bureau 1998) to over 15m in 2006 (but with a fall to about 11.5m in 2001 and 2002). On the other hand, domestic visitors went from 13.4 million in 1996 (London Tourist Board and Convention Bureau 1998) to a peak of 18.5 million in 2000, dropping to 10.9 million in 2006. There are several reasons for this change. Firstly, the global tourism market is currently undergoing expansion, with total arrivals rising from 687 million in 2000, to 842 million in 2006 (UNWTO figures, cited in VisitLondon 2007). This means that, despite the apparent growth in overseas visits, London's market share of international tourists compared to other destinations is actually decreasing (London Development Agency 2004). Secondly, the under-performance of London in the domestic market (London Development Agency 2004) seems to be due to the high cost of visiting London and the continuing availability of cheap flights to the continent.

Besides the two traditional visitors categories (domestic and overseas), it is interesting to note that in a large metropolis such as London, a third type of tourist, the 'internal' or 'local' tourist', needs to be considered. In 2004 for example, 9 percent of London's domestic overnight visitors were from London itself (VisitLondon 2005b). Londoners also account for more than 70 percent of the total number of day visits to other parts of London (Countryside Commission 2005 and VisitLondon 2005b). Internal tourists are therefore an important ingredient in the tourism market. Their income is higher than the national average and they spend a much higher proportion of it on leisure related goods. According to a study conducted by the

Greater London Authority (GLA Economics 2003), Londoners' expenditure on restaurant and café meals is 40 percent above the national average and, on sports and leisure, cinemas, theatre and other entertainments, it is 30 percent higher (GLA Economics 2003 and VisitLondon 2005a). In addition, internal tourism has the advantage over other types of tourism as it is an all-year round market with very small seasonal fluctuations (VisitLondon 2005a).

Purpose of visit figures (VisitLondon 2007) show that leisure tourism is the dominant market, with over 50 percent domestic and 42 percent international tourists visiting London mainly for leisure reasons. However, the Visiting Friends and Relations (VFR) market accounts for about 20 percent of domestic and 23 percent of overseas visits, and should not be overlooked. In particular the domestic VFR market appears to be growing in value, with an average spend increasing from £70 per visit in 2000 to £93 in 2006 (VisitLondon 2007).

Business tourism takes a similar share to the VFR market (27.1 percent of domestic and 23.5 percent of overseas visits), but it accounts for almost 30 percent of the total tourism spend (VisitLondon 2007). The business market's value is in fact more than fourfold the one of VFR: the average spend per visit in 2006 was £293 for domestic business visits (compared to £203 leisure and £93 VFR) and £690 for overseas (compared to £436 leisure and £361 VFR) (VisitLondon 2007).

Despite being one of the leading business destinations in the world, London today does not rank among the first international convention destinations, falling from number one in the late 1970s to sixteenth in 2002 (London ICC 2005a). In 2002, for instance, London was only eleventh in the European league table of market share of meetings, overtaken not only by world cities such as Paris and Berlin, but also by smaller destinations like Helsinki and Edinburgh (London ICC 2005b). The reason for this seems to be the lack of adequate conference spaces and especially of a central, multi-purpose conference centre (London ICC 2005a). Paris Expo, for example, with its 212,000 square metres of floor space, is larger than all the exhibition venues in London combined and overall the French capital has more exhibition space than the whole of Britain (Cotton 2003).

GOVERNANCE AND TOURISM POLICY

Despite the leading role of London as a world tourist destination for over a century, it was only in the 1990s that tourism became recognised as an important aspect of London's economy and became part of the City's development policies (London Tourist Board and Convention Bureau 1997). The 1985 government's report *Pleasure, Leisure and Jobs in the Business of Tourism* (Department of Employment 1985) was the first document to highlight the important contribution of the tourism sector to the British economy

(BTA 1988). Soon after, in 1987, the Joint London Tourism Forum developed the first Tourism Strategy and Action Plan; tourism strategies continue to be published every four years by VisitLondon.

This new awareness of the importance of tourism did not occur in a particularly prosperous period. Between 1990 and 1992 the economic recession in the UK brought about a decline of more than 1.5 million visits (Bull and Church 2001). Nevertheless, in the following years, London's tourism witnessed an astounding recovery, with its total visits rising from 15.7 million in 1991, to 26.2 million in 1996 (Bull and Church 2001). An important factor leading to this increase was the weakening of the pound following the UK's departure from the European Community's Exchange Mechanism in 1992 (Bull and Church 2001).

It is interesting to note that, at that time, the motivations to visit London were fairly different from the ones of today. In 1995, for example, the main purpose for domestic visitors was visiting friends and relatives (46 percent of the total, more than twice that in 2006), whereas holiday was a main motivation for only 31.7 percent of domestic visitors (Bull and Church 2001) compared with 50.2 percent in 2006 (VisitLondon 2007).

In 1996, the Government Department for Culture Media and Sport (DCMS) funded the development of a London brand and a marketing campaign (London Tourist Board and Convention Bureau 1997). However, the real turning point for the capital's tourism governance and policy was in 2000 when, after twelve years without a London-wide authority, a directly elected mayor and assembly were created. In 2001 the Mayor delegated his responsibility for tourism to the London Development Agency, which conducted a review of tourism activities in London and, following the publication of a new Tourism Strategy, created VisitLondon (a public-private partnership responsible for all marketing activities). In addition, around the new millennium, 2000 London opened a number of new attractions that are now among the most visited in the UK: Tate Modern, Millennium Bridge, London Eye, Millennium Dome and 30 St Mary Axe Tower. These changes in London's governance and physical structure, together with investment in new attractions, delivered a significant peak in visits that rose from 28.7 million in 1999 to 31.6 million in 2000 (VisitLondon 2004). However, these numbers were not sustained and in 2001 a number of threats harmed London's tourism sector, causing a drop of 5 million visits, that so far have not been regained.

The crisis originated in February 2001 when the foot-and-mouth disease spread in the British countryside and images of burning cows were broadcasted around the globe (Hopper 2003). It then reached its highest point after the 9/11 terrorist attacks in the US, and continued into the following years with the outbreak of SARS in 2003 and the London bombings in July 2005.

To respond to these threats, in September 2001 the London Tourism Recovery Group, and later the same year the London Tourism Action Group, were established (Hopper 2003 and London Development Agency

2004). By the end of 2001 a rescue plan was also developed, that involved a number of marketing actions: first, a campaign to promote London theatres, then from January 2002 to April 2002 the 'The greatest show on Earth' campaign, and finally, in February 2002, a second phase that focused on special deals on hotels, attractions and restaurants. Despite these marketing campaigns, visits to London continued to decrease between 2001 and 2003 (VisitLondon 2007). As a result VisitLondon, realising for the first time the importance of internal visitors, in 2003 launched a *Totally London* campaign aimed at enhancing civic pride and encouraging Londoners to experience London's facilities and attractions. In 2005, VisitLondon also produced *London for Londoners* (VisitLondon 2005a), a practical guide for London boroughs on how to promote tourism to locals. Today, the four main principles on which tourism policies are being developed (mostly by the London Development Agency through VisitLondon) are growth, dispersal, resources and diversity and inclusion.

The two key markets on which London is now focusing are the business and the VFR markets, being smaller than the leisure one and with a greater potential for growth. In particular, London aims at improving perceptions of value for money as the high cost of visiting London is one of the main reasons for the decrease of domestic trips and stability of overseas visits. In addition, improving the quality standards of the visitors' experience, especially accommodation, information and transport, constitutes a fundamental means of raising cost/quality perceptions (London Development Agency 2006). In order to expand the business market, the construction of an international convention centre is also being reassessed, with various feasibility studies having been carried out in the past fifteen years (for instance London ICC 2005a).

According to London authorities, the upcoming 2012 Olympics will be a unique chance to promote London to be the number one destination worldwide, allowing a lower dependency on few key markets like the US and western Europe. The primary aim of the games, however, is not to increase absolute numbers, but rather to spread the benefits of London tourism, especially to the least developed areas in the east and south-east of the City and to the rest of the UK thanks to London's role as a gateway to other regions (Smith 2006).

TOURISM DISTRIBUTION AND OPENING UP NEW AREAS

Spreading the economic benefits of tourism throughout the City represents a fundamental policy objective in London and it appears among the first aims of every strategy document or action plan developed by the London Development Agency (see for instance London Development Agency 2002, 2004 and 2006). Central London remains the predominant area for tourist accommodation, with over 56 percent of hotels being located in the three

Boroughs of Westminster, Kensington and Chelsea and Camden (VisitLondon 2007). Most of London's best known tourist attractions, including twelve of the UK's top visitor attractions, are located within the Circle Line of London Underground (London Development Agency 2002). Visits to attractions in Central London every year are estimated at 36 million, whereas analogous figures for other London sub-regions are remarkably lower: 2.8 million for South London, 5.4 million for East London and 0.4 for North London (London Councils 2006).

The potential for development of new tourism areas ought to be large, not least that London is affected by a fundamental problem of under-supply of accommodation, with resulting high occupancy and prices. Potential for drawing tourists from the centre also comes from the rapid growth of global tourism and the emergence of a new type of visitor: an experienced traveller who is eager to flee the traditional 'tourist bubbles' in order to explore new, 'unspoilt' destinations (as discussed in Chapter 1). These contemporary tourists, described by some commentators as 'adventurers' (Eade 2002), are characterised by being more independent than the traditional visitor (Poon 1993, Eade 2002 and Maitland 2006) and not satisfied with the mass tourism offer, rather seeking 'authentic' and 'non-homogenised' experiences (Maitland 2006). In contrast with Feifer's 'mass post-tourist' (Feifer 1986 and Urry 1990), another key characteristic of this 'new tourist' (Poon 1993) is the search for authenticity or, as Getz (1994) called it, the genuine, the unadulterated, the 'real thing'.

This trend appears to be especially important for an established tourism destination such as London, where a survey conducted by the London Development Agency in 2007 revealed that 44 percent of overseas, and 85 percent of domestic, overnight visitors had stayed in London at least twice in the past five years, and 81 percent of domestic day trippers, in the same period of time, had visited London at least ten times (London Development Agency, London Visitor Survey Jan–March 2007, (DA 200)).

This idea of a new, experienced type of tourist brings us back to the importance of internal tourism in London and to the fact that tourists, workers' and residents' demands in terms of places to work and play are becoming increasingly alike (Maitland 2006 and 2007). As we shall see in the studies in the second part of this chapter, of two new tourist destinations in London, Islington and Bankside, contemporary city users appear to be attracted by a sense of distinctiveness or 'placefulness', and a set of qualitative assets connected with the City's quality of life and atmosphere.

Well established in London is the process of touristification of ethnic quarters where there is a quest for the 'exotic', and in a city of great ethnic diversity there are many potential opportunities to experience the 'authentic London experience' Shaw Bagwell *et al.* (2004), for example, in their longitudinal study of Brick Lane (Spitalfields, Tower Hamlets) and Green Street (West Ham, Newham) have shown the development of these two East London multicultural neighbourhoods from gritty but characterful areas to new destinations for leisure and tourism. The branding of 'Little India' in

Southall, west London, also reflects the ability of ethnic markets to draw visitors from beyond the local area.

The attraction of visitors – Londoners and international visitors – has also developed hand in hand with some of the City's new 'cultural quarters'. The creative clusters in Hoxton and Shoreditch that developed rapidly during the 1990s drew large numbers of visitors to the range of hip galleries, bars and night clubs. Other areas, for example Deptford in south London, aim to replicate the Hoxton experience as local policy makers take up the London Development Agency's call to disperse visitors from the centre and encourage them off the beaten track.

In the next part of this chapter we examine two 'new tourism areas' in greater depth. London continues to draw large numbers of visitors and, as we have seen, in recent years international visitors have replaced domestic tourists. The City's central area has been revamped with some new attractions, the upgrading of existing institutions – for example Tate Britain and London's Opera Houses – and new investment in public transport and security. Tourism policy was boosted by the arrival of mayoral government in 2000, and it has had a consistent ambition to diversify London's offer and encourage the opening up of new tourism areas. We examine the role of tourism in the recent transformation of Islington and then similar processes at work behind the more visible transformation of Bankside (see Figure 4.1).

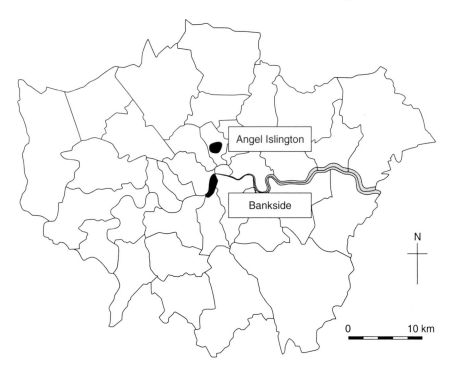

Figure 4.1 London: Islington and Bankside areas
Source: Authors' own

ISLINGTON – THE 'REAL LONDON'

The London Borough of Islington is part of inner London and stretches from the edge of the central area northward towards the suburbs. Like much of London, it consists of a series of comparatively distinct neighbourhoods that developed from original villages. Tourism has been focused at the southern end of the Borough, closest to central London in the districts of the Angel/ Upper Street, and Clerkenwell. The Angel is comparatively close to King's Cross/St Pancras, a major transport interchange although situated in what has for long been a rundown area. Clerkenwell abuts 'the City', London's main financial district, and stretches north toward the Angel. These districts are the heart of our discussion. Over the last quarter century they have experienced a process of radical change and development that has transformed their resident populations, economic activities and image. Much housing has been gentrified, former industrial buildings have been converted to private housing or service industries and the area has become very fashionable. It has also developed as a tourism destination. In 1980 it attracted a negligible number of visitors, but by 1998 there were estimated to be some 4.5 million visitors per annum, spending some £105 million, and accounting for 4,700 jobs (Carpenter 1999). Like most estimates of tourism numbers, these figures need to be treated with some caution, but the magnitudes seem to be widely agreed.

The argument in this section is that this notable growth in tourism has been part of a wider process of development and gentrification, both contributing to and reflecting change in London. We begin by surveying Islington's tourist products and how they have developed, and the role of policy in that process. We then draw on original research from visitor surveys and interviews with visitors – for example Maitland and Newman (2004) and Maitland (2007a) – to consider the appeal of the area to visitors, and suggest that there are strong links between the demands of [some] visitors and [some] residents of the area.

In one sense, tourism and leisure in Islington have a long history. The area was one of the playgrounds of pre-industrial London, and there is a legacy in its numerous (small) theatres and pubs. However, as we have seen, London's attitude to tourism has, until comparatively recently, been unenthusiastic, despite the numbers of visitors it attracts. In the late 1970s, the Greater London Council (the then City government) emphasised the industry's low pay and inconvenient working hours before conceding somewhat grudgingly that it offered employment to 'poorer families and to workers whose employment prospects have suffered so much in London's population decline' (Lipscomb and Weatheritt 1977, p. 17). Changes in attitudes to tourism were driven by rapid economic change and decline in many traditional industries in the late 1970s and 1980s in London. Tourism was seen as one of the few possible growth areas in inner cities and rapidly became important in the emerging 'inner city' and later 'regeneration' agenda – see for example ETB (1981) and

Beioley, Maitland *et al.* (1990). Policy in Islington reflected this shift of perspective. The first policy attention to tourism came in the 1980s, through the Inner City Programme, a central government initiative designed to channel resources to areas hard hit by economic change. Islington incorporated tourism work in its economic programmes from 1985 (Discover Islington 1992).

Islington, and particularly Angel/Upper Street and Clerkenwell could be seen to have tourism advantages. They are highly accessible to central London, mainly through underground and overground lines and by bus, although services are often crowded and suffer delay. Angel/Upper Street are within walking distance of King's Cross but the walk is made unappealing by heavy traffic and the area around King's Cross remains intimidating, especially at night. This may change following the re-opening of St Pancras station as the Eurostar terminal in November 2007, and the transformation of the station into an eating and shopping destination in its own right. Day visitors and Londoners looking for an evening out can get to the area easily, as can overnight visitors staying in the hotels of central London.

What would attract them? A review of Islington's tourism product was carried out in 1992 as part of the work on a Tourism Development Action Programme for the area (Discover Islington 1992). The review summarised the pull of the area, drawing on the developmental work of the 1980s. It had two elements: first there are visitor attractions of a familiar and conventional sort, housed in specific buildings: arts venues such as Sadler's Wells (opera and dance), the Almeida Theatre, music venues like The Rocket, exhibition and conference venues – notably the Business Design Centre – small museums; and Arsenal football club's stadium. Second, there are much smaller attractions few of which would bring in visitors in their own right, but that as a cluster could create appeal. These included markets, restaurants and bars – many individually run rather than part of chains, and some highly rated in restaurant guides – speciality shops – including Camden Passage Antiques Market and designer shops in Upper Street – and art and craft galleries. The review also drew attention to the area's 'rich architectural heritage . . . the squares of Amwell, Clerkenwell, Barnsbury and Canonbury are exceptional . . . rich in historical associations' (ibid. 9). However this was offset by it 'not necessarily [being] viewed as a tourist destination, dirty/shabby image, crime rate/fears for safety' (12). The review summarised Islington's product and market as 'promoted to date as the "Real London" . . . a special and high quality place to visit for those who don't like to think of themselves as tourists' (22) and argued that this focus should continue. As pointed out above, by the end of the 1990s, Islington was well established as a tourism destination. However, paradoxically, this success was not primarily due to tourism policy.

There are two reasons for taking this view. First, the substantial public and private investment that occurred and that led to improved attractions and tourism infrastructure was not directed at developing tourism. Second, the major appeal of the area for visitors turns out not to be conventional tourist attractions, but rather qualities of place and consumption in an area perceived to be off the beaten track. Both place and consumption opportunities have been radically upgraded, but this was not driven primarily by tourism policy or with the intent of developing tourism. We examine these points in turn.

A series of investments brought improvements to Islington's tourism infrastructure and attractions from the 1980s onwards.

- The main underground station was reconstructed and transformed from an intimidating and unattractive station with limited passenger capacity to a modern and welcoming facility;
- regeneration strategies around King's Cross and the City fringe made the approaches to the area more welcoming and reinforced the importance of tourism (Long 1999);
- new leisure attractions were created or substantially refurbished – for example the Crafts Council, Sadler's Wells Theatre and Almeida Theatre;
- the Business Design Centre (BDC), a major new business exhibition centre, was developed by local entrepreneurs, attracting business visitors;
- new hotels opened in the late 1990s – for example Hilton hotel adjacent to the BDC, and Jury's hotel.

In addition, the area benefited from organic brand building. Islington was the home of Tony Blair, elected Prime Minister in 1997, and a popular figure in the run up to the election and in the years immediately afterwards. His high profile and association with fashionable Islington (see below) generated considerable coverage for the area in the media and numerous mentions in guide books.

Although Islington's tourist offer was being improved however, this was not primarily part of a tourism strategy, and the changes were not aimed at tourists. Discover Islington, the public–private local tourism organisation did some skilful work in positioning the area and encouraging visitors, but was closed in 2000. Improvements to the underground station were to a long planned scheme to increase capacity at a station that was often dangerously overcrowded at peak commuting times. Regeneration strategies in King's Cross were aimed at improving the image and functioning of an area well known for drug dealing, petty crime and prostitution. The BDC was an economic initiative that found a new use for a listed building and was supported under the then Urban Development Grant scheme. Substantial investments in arts venues were funded by the National Lottery from funds dispensed on the basis of artistic quality rather than location, and with the intention of benefiting a domestic rather than a tourist audience. Blair brand building was, obviously, not addressed at tourists. The hotel developments clearly owed something to the recognition by the 1990s that the area had become a tourism destination – but also to the search for lower cost sites with good access to central London for budget or mid-range hotels.

It is not clear either that conventional visitor attractions constitute the main appeal of the area for tourists. Visitor perceptions were investigated through a visitor survey and subsequent in-depth interviews. The survey questioned overseas tourists visiting Islington. They were asked a series of demographic questions: how they had found out about the area, and what they thought of it. To allow some element of comparability with visitors to London as a whole, questions were designed as far as possible to be compatible with the Survey of Overseas Visitors to London, a sample survey carried out annually by London's tourism organisation (formerly London Tourist Board and Convention Bureau, now VisitLondon). The findings have been reported more fully elsewhere (Maitland 2003 and Maitland and Newman 2004). It turned out that visitors to Islington differed from overseas visitors to London as a whole, tending to be older, to have visited London previously, and to be linked to the City through friendship or family networks. They were not drawn to the area by conventional visitor attractions rather, they sought and enjoyed the broader qualities of place – a physical environment in which architecture, building, streetscape and physical form, combined with socio-cultural attributes such as atmosphere, and a perception of the area as 'cosmopolitan' and 'not touristy'. Whilst shops, bars and restaurants were valued, it was as part of a broad landscape of consumption adding to the quality of place – particular shops or restaurants were not attractions.

For these respondents, the dirty, shabby, unsafe image of Islington of the 1980s and early 1990s seems to have been banished, and this perspective was borne out by in-depth interviews (below). Since the appeal of the area derives from positive place qualities and the landscape of consumption, we need to examine how these have been created. This means that we need to consider the wider processes that have driven change in Islington, particularly the process of gentrification. As Butler (2003) points out, many of the middle-class in London are 'cosmopolitans living in a metropolitan environment', and although 'in boroughs such as Islington the middle class is never a numerical majority, [its] visibility and influence . . . far outstrip its physical presence' and being a minority 'has not prevented its members from defining such areas in their own image' (Butler 2003, p. 2470). Amongst the effects of gentrification have been renovation of buildings and improvements to the physical environment: 'beautifully renovated houses and cottages line ordered and generally well cared for streets' whilst 'council estates are screened off' (Butler 2003, p. 2474). At the same time, affluence has transformed the consumptionscape. Upper Street 'where restaurants, themed bars, kitchen/bathroom shops and estate agents have edged out retailers of a past era . . . has become a global space, servicing the international service-class diaspora in an environment that acknowledges the cultural capital of the customer' (Butler 2003, p. 2474). For those occupying gentrified housing, much of the appeal of the area lies in feeling that they enjoy both diversity and cohesion – the architecture and feel of an old place combined with a buzzy metropolitan environment. Whilst in Upper Street social interactions may be limited, and in reality it increasingly caters for the very well off, it has 'an aura of inclusivity and this is a large part of its appeal' (Butler 2003, p. 2476).

How do these changes fit with visitors' perspectives? A series of interviews with visitors allowed in-depth exploration of their perceptions of the area (and of Bankside, see below). Visitors from a variety of countries around the world were interviewed, most of whom turned out to be in professional occupations (though that was not the basis of their selection) (Maitland 2007 and 2008). Quotations from the interviews are shown in italics. Much of what visitors said revolved around architecture and sense of place, consumption opportunities and interaction with 'local people'. The architecture and older buildings were frequently mentioned – *the architecture, the houses are old, and they are really nice* and interviewees often discussed particular small aspects of buildings: detailing of windows, lintels, *the chimney stack on a building down that alley*. The architecture and streets contributed to the feeling that this was a *high class community* and one that is *a lot cleaner, nice and tidier than other parts of London* and *safe* to walk around. For some the area was *pricey* though still an interesting place to shop (see Figure 4.2). Consumption

Figure 4.2 London: Islington
Source: Bill Enckson

opportunities could be enjoyed in a historic environment: [those] *quaint little shop fronts are quite deceiving . . . you look inside and think, woah, that's a beautiful shop in there.* There are *very nice restaurants and bars* although, compared to much of central London, it was *a quieter part of the City.*

For visitors, Islington was not a 'touristy' area, despite the presence of themselves and other visitors. The presence of local people was important so *you are not treated as a tourist* and *can look out on the area at the people, looking at how they really live.* It was easier to meet people because *people are more relaxed, they sit down, drink coffee, they talk a lot . . . because it is more relaxed.* The area is like *a little arty community* and these respondents felt *you are not actually an outsider . . . you are not treated as a nobody they have a little conversation with you.* Even the most mundane features of everyday life were of interest in this context, including a visit to the local branch of Tesco, a mid-range supermarket. One respondent recalled *I went to Tesco to buy things, well it was an incredible experience because how people buy their food, the people wear different type of clothes,* whilst another remarked *One of our favourite places has been Tesco . . . we just like to look around and see.*

It is clear that these visitors were experiencing the aura of inclusivity that Butler notes, and that they were enjoying the environment and consumption opportunities that gentrification had brought. Conventional visitor attractions – Sadler's Wells, the BDC, the Almeida – rarely figured in their accounts of Islington's appeal, whilst the sense of being in an area that was not 'touristy' and where it was possible to see and do what the locals do was very important. The visitors were professionals, frequently experienced travellers (as were most of the respondents to the visitor survey) and so it is unsurprising that they fitted in well with Islington's 'cosmopolitans living in a metropolitan environment'. They too can be seen as members of the groups variously described as cultural class international élites (Rofe 2003) or 'cosmopolitan consuming class' (Fainstein, Hoffman *et al.* 2003), enjoying the 'mixture of the "traditional" and the "exotic" demanded by a new urban intelligentsia' (May 1996, p. 200). They too can be seen as enjoying a 'fantasy of place' (May 1996, p. 203) that appears to offer both diversity and cohesion, whilst catering almost exclusively for global élites.

The story of tourism development in Islington is thus a complex one, and three particular points deserve emphasis. First, tourism developed without major public investment that was focused on tourism. The improvement in the attractions of the area was contingent on a number of other investment strategies and policies that were not linked to tourism – upgrading the Underground, and a major injection of funding into the arts from the National Lottery, for example. Tourism policy had only a very limited role to play. Second, the appeal of the area to overseas visitors is in any case not based around conventional tourism attractions, but about opportunities to enjoy consumption in an area that is architecturally and environmentally attractive, and that is obviously also used by 'local people' who mark it out as distinctive and in some sense authentic (Maitland 2007b). In this sense, the absence of planned tourism development has added to the attraction of the area since it means it does not seem touristy: instead, as one visitor said *it's like something that has evolved: it's not been someone's plan*. Gentrification has helped create an area that visitors to Islington enjoy; and the visitors themselves, of similar class and taste, have helped drive along the process. This takes us to our third point: the importance of mingling with 'local people' for the visitor experience suggests that the decision to position Islington as somewhere for tourists to find 'the real London' was an astute one. However, we must remember that the Islington enjoyed by visitors is the 'real London' only of the cosmopolitan consuming class. It appeals to those who are members of that class, whether they live in the area, are temporarily resident in London, or are there on a visit (often a repeat visit). It does not include the London of poorer residents, whose shopping and leisure activities increasingly take place not in Angel/Upper Street, but in the downmarket Caledonian Road – nearby but effectively screened.

REINVENTING BANKSIDE

By the end of the 1990s Bankside enjoyed a global reputation as a new tourism destination. But as in the case of Islington we argue that this has not been the outcome of a coordinated tourism strategy and that the current appeal of the area owes as much to visitors' appreciation of local character as to investment in new attractions.

Bankside now forms part of a South Bank extending from Westminster Bridge in the west, to Butler's Wharf in the heart of London's docklands. The riverside walk, only broken by the difficult crossing of London Bridge, adds to the impression of a seamless tourism quarter linking newer attractions such as the London Aquarium in the refurbished County Hall with the iconic Tower Bridge. Bankside, the part of the South Bank between Blackfriars and London Bridges, offers itself as a model of tourism and culture-led regeneration. The rapid physical transformation of the area during the 1990s was an exemplar for many other cities in search of a new approach to urban renewal. Not least among those eager to learn were civic groups in New York looking at possibilities for the rebuilding of Lower Manhattan (McKinsey & Co. 2002). What culture and tourism seemed to offer was the capacity to attract commercial land uses, jobs and rising property prices. As Bankside became a model for other cities, at the same time it had messages for tourism policy in London. After all, what Bankside showed clearly was that tourists could be persuaded to cross the river, leaving the established heritage and shopping attractions on the north bank, and then gradually move eastwards. The redevelopment of docklands in the 1980s signalled national policy priority to east London, subsequently reinforced in regional plans and prioritised by the new mayor. London's riverside is now perceived as a major asset and opportunity for the sorts of development that will attract and accommodate the urban living preferences of the middle-class that supports the City's competitive global economy.

The redevelopment of Bankside over the past twenty years is seen as a success story. The Bankside model is important to discussion about the future of tourism in the City and in particular to the broad policy objective of using tourism in the regeneration of east London.

However, the lessons from Bankside may not be so straightforward. We need to take a more critical look at the transformation of this part of London. Studies of visitors emphasise the less tangible attractions of the area over the higher profile attractions such as the Tate Modern Gallery. Surveys by VisitLondon (2005) on the wider South Bank area suggest that, despite a range of attractions, the most popular activity amongst visitors surveyed was 'going for a stroll', followed by visiting a restaurant, café, pub or bar. Whilst it seemed that Tate Modern could act as a primary motivation to visit, the area is 'used heavily as an informal place to stroll' and 'levels of visiting the main attractions are low' suggesting that they were not drawing visitors to the area.

More detailed investigation of foreign visitor attitudes through in-depth interviews confirm and flesh out these findings (Maitland 2007b and 2008). The centrality of the area, its place between established tourism attractions and the convenience of the Millennium Bridge as a new river crossing has meant that it has been readily incorporated into sightseeing circuits. *I started from Westminster Abbey, then crossing (*sic*) Westminster Bridge . . . it's very fascinating to walk alongside the Thames river . . . there's St Paul's cathedral.* The riverside walk allows easy access to familiar landmarks that retain resonance for visitors – for example, a young Indian couple had walked along to London Bridge, as they had both learnt the nursery rhyme 'London Bridge is Falling Down' as children. Walking is key to the way visitors explore and perceive the area, and is mentioned frequently (as it is in Islington interviews, above): *we walked fairly far; a good place to walk around.* In their strolling around, visitors appraised the architecture and design of the Bankside area. Whilst some saw it in broad terms as a pleasant back-drop, reflecting London's history and heritage – *we like to walk in nice places* – others spoke in detail about the architectural and physical character of the area, and the contrast between old and new buildings. Strolls might be unplanned – *we just get off the Underground and walk around* – but some visitors could be engaged with and perceptive about what they were seeing.

For visitors, Bankside was clearly different from London's established tourism precincts, and it was frequently described as *not touristy* or *out of the way* or *off-the-beaten-path.* Interviewees contrasted the area with other parts of London they had visited: *very commercial* places like Piccadilly Circus, or crowded areas – *Buckingham Palace and about 10,000 people.* They also drew contrasts with parts of other capitals – *if you're in Paris on the Champs Elysées, you feel there are only tourists.* The sense that the area was not a tourist zone was true even for visitors whose circuits had included what tourism marketers would undoubtedly see as Bankside's major visitor attractions. One interviewee who had progressed from Tower Bridge and Southwark Cathedral, inspected the Clink prison and the *Golden Hind* replica ship, and was intending to see Shakespeare's Globe and visit Tate Modern was nonetheless clear that they were not in a touristy area. The solution to this apparent paradox seems to lie in the mix of activities and users of the area. Visitors noted that Bankside is an area where people live and work and thought it *nice that you can live close where you work – it's not like that in Düsseldorf* (interviewee's home city), whilst being aware that to enjoy this *you've got to be rich . . . more than rich.* The sight of people at work in offices contributed to the tourism experience. For one couple, evidence that they were experiencing *the real London* came from being able to look in office windows and see people typing away at computers – *it's really cool.*

The presence of people who were not tourists, but Londoners working, shopping and relaxing acted as a marker indicating that this was an area in which the 'real' city could be experienced. In a place where *there are many people from London* the tourist experience is affected: *you don't feel like you're in Disneyland, you're in a city where people actually live, it doesn't feel artificial, as if it was made for you.* Bankside has of course been 'made' in the sense a coherent design strategy has been applied to its development (Teedon 2001). However, the point is that visitors do not feel it has been made for tourists, but rather that it is an ordinary – if gentrified and upmarket – part of the City where everyday life goes on for a mixture of people. This echoes comments by visitors to Islington who felt the area had evolved and had not been part of a plan. That makes it is more interesting: *it's just more authentic and it's just fun because local people and tourists, they also mix.* In the minds of visitors, the area is not dominated by its new, 'iconic' tourism attractions or made a 'tourist area' by its heritage. VisitLondon found that the 'impression among many overseas visitors [is] that the South Bank/Bankside area is somewhat "off the beaten track"' (VisitLondon 2005b, p. 89). Deeper exploration of visitor attitudes show that at least some visitors look attentively at the area and those who use it, and judge that they are doing what the locals do, and thus they have left tourist zones, and are enjoying 'the real London'.

The qualities highlighted by visitors give an emphasis to consumption over other visitor attractions. If, as these findings suggest, visitors are drawn to consumption, the character of the area and less tangible qualities, then the success of Bankside is a rather more complex story than one of simple investment in cultural buildings to attract tourist spending. An important part of this story will be questions about the role of the public policy in creating new places and revitalising cities. London policy makers welcome the transformation of the riverside but we need to consider how far they were able to influence this process. Certainly visitors have been drawn away from main tourist routes, but how and why this happens needs close examination. It seems unlikely that there can be easy replication of the Bankside process in other areas.

The recent history of the South Bank in central London has been controversial and the loss of port activity with its industrial infrastructure and manual jobs has been bound up with political transformation. There are not surprisingly different claims about the reasons for successful urban renewal. In the 1970s, community groups struggled with local and national governments over the future of the stretch of the South Bank between Waterloo and Blackfriars Bridges (Brindley, Rydin *et al.* 1996). The new restaurants, shopping and open spaces reflect that community victory and a failure of less sensitive public policy. Politics, rather than planning, opened up the South Bank to visitors. Blackfriars Bridge marks an administrative boundary between London boroughs and, on the Southwark side, there were different struggles between local community, commercial developers and a borough council undergoing a substantial change of image and direction throughout the 1980s and 1990s. One account of the Bankside story views transformation of the area through the lens of innovative public policy and the roles of local actors in making the case for tourism development (Tyler 1998). Others see the process as less clear cut, and view the contingent co-location of cultural facilities to have been largely beyond local control (Newman and Smith 2000)

According to Newman and Smith, the forces driving change on Bankside were a dynamic property market and decisions taken by central government, notably the extension of the Jubilee line with a new station at Southwark, and the allocation of substantial funds from the new National Lottery to large arts projects. Central government also provided the funding for an improved pedestrian link across the river at Hungerford Bridge, and the new Millennium Bridge (creating a new route from St Paul's to Tate Modern) that were opened in 2002. In the 1990s, some new cultural attractions opened, such as the replica Globe Theatre, and the Tate Modern Gallery exceeded its original estimate of three million visitors a year. A series of minor attractions lined the route to London Bridge where environmental improvements encouraged more foot traffic through the area. However, it was only after these major investment decisions had been taken, and not in any coordinated way, that Bankside began to be treated as a coherent area in local tourism policy.

Whereas in the 1970s and 1980s the Borough of Southwark had largely failed to influence development in the area (McCarthy 1996) it was the new forms of government that emerged in the 1990s that began to have a coordinated impact on Bankside. New partnership bodies, the Pool of London Partnership, Cross River Partnership and the Better Bankside Business Improvement District (BID), produced more coherent tourism and cultural development strategies and secured government funding for environmental improvements. The Bankside BID grew out of a less formal 'marketing partnership' of local businesses and has straightforward objectives of attracting more customers and making staff, visitors and clients happier to come and do business in the area. The BID demonstrates a clear understanding of the importance of the shared interests of visitors and businesses. Such views have the support of a borough council that has come a long way since its opposition to tourism development before the 1990s. Political control of Southwark Council changed in the 1990s and Southwark now sees tourism as a key industry evidenced by its employment of full-time tourism officers since 1996.

Attitudes to the development of Bankside changed and local government and the new partnerships were able to contribute to the rebranding of the area from the mid 1990s through marketing a coherent destination and through signage and routing (Teedon 2001). The name Bankside was given a new meaning, taking in a wider area than in the past, and quickly establishing itself in both tourism strategies and visitors' vocabularies.

However, whilst public bodies developed a strong role in marketing, the borough council had rather less control over the residential, hotel and office developers drawn to this section of the London riverside. We could see the planners as being swept along by the tide of riverside development in London's 'riverside renaissance' (Davidson and Lees 2005). A return of the middle-class to the centre of the City was encouraged by government through the Urban Task Force (DETR 1999) and subsequent Urban White Paper (DETR 2000). This emphasis on gentrification was followed up in the London Plan (Mayor of London 2004). Planning policy has supported the social change that has been marked in areas near to the river. According to Davidson and Lees, the population of managers and professional groups living along Southwark's riverside grew by over 200 percent between 1991 and 2001 compared to an increase of 76 percent inland, a rate of change broadly comparable to that across inner London as a whole (Davidson and Lees 2005 1177). Riverside locations hold particular attraction for a new urban middle-class. In north Southwark new restaurants and shops accompanied this change and London's remaining central fruit and vegetable market, at the eastern edge of Bankside, has experienced its own renaissance. Redevelopment projects were resisted in the 1970s and 1980s and Borough Market now supports high value food stalls and attracts residents, Londoners and visitors.

Problems associated with gentrification elsewhere in London, the displacement of working-class residents for example, have been less marked along the riverside where old commercial buildings offered opportunities for a development-led style of gentrification. However, new residents and visitors to Bankside dominate public spaces and have their impact in the upscaling of local consumption opportunities. The transformation of Bankside suits visitors and new residents alike.

As we have seen, for many visitors, it is consumption, character and less tangible qualities that combine to make this a successful urban quarter. We suggest that the process of change has not been planned, and the lessons for tourism regeneration elsewhere in London need careful consideration.

Bankside cannot be replicated elsewhere. We certainly need to see this new tourism area in the context of social and economic change in London as a whole and in particular the transformation of the riverside. The gentrification of the riverside goes hand in hand with the creation of cultural attractions and upgrading of local services and environmental quality.

CONCLUSIONS

There are clear similarities in the processes of urban renewal at work in Islington and Bankside. The changing economy has created new middle-class preferences for urban living and preferences shared between residents and visitors have driven urban change over a period of fifteen years or so. Understanding the attraction of visitors to these desirable locations has to be set in the broader understanding of urban change and the process of urban gentrification in particular. Can this process be replicated in other parts of the City? If gentrification continues to drive urban processes in the same way then the answer is probably yes. Both Islington and Bankside needed some upgrading to connections by tube to make it easier for visitors to discover these new places, and the longer term investment in London's underground network and stations may enhance tourism potential elsewhere. More broadly, such urban change depends on the continued growth of London's business centre and the finance to support the high price of residential renewal.

In both cases, the process of urban change was achieved with relatively little conflict. Despite the displacement effects noted by gentrification research, many studies of gentrified neighbourhoods emphasise the tolerance of different groups sharing the same urban spaces. In Notting Hill, for example, working-class residents were preoccupied with their daily lives and did not share, or come into conflict with, the aesthetic values that the new middle-class imposed on the area (Martin 2005). In the 1990s Notting Hill became a draw for visitors to London, keen to experience the locations of the film of that name. We might contrast that experience with the dispute about the filming of the book, *Brick Lane*. Neighbourhood politics can clearly be more controversial in some localities. In Spitalfields, in east London, ethnic diversity has been exploited through business development, the use of festivals and events and rebranding. However, the middle-class conviviality that characterised Islington and Bankside is only one storyline and where a new tourism economy meets areas of deprivation and discrimination conflict may accompany conviviality.

The rebranding of the East End has to be seen alongside a rebranding of the City and country as diverse and tolerant. On winning the Olympic Games, the Prime Minister said: 'London is an open, multi-racial, multi-religious, multicultural City and rather proud of it. People of all races and nationalities mix in with each other and mix in with each other well' (*Guardian* 2005).

London's diversity has become an important part of its marketing, but needs government at all scales to maintain the image of multiculturalism that 'now serves as Britain's distinctive rationale in the current world order, and in many respects can be considered a success' (Dench, Gavron *et al.* 2006, p. 226). The local interactions between city users in Brick Lane are important for national as well as local politics, and the success of tourism policy in London in drawing visitors from the traditional attractions, and into the diverse inner city, depends on policy support from higher levels. The Mayor's cultural policy gives priority to diversity, but also recognises the importance of the City's 'world class' assets (Mayor of London 2004b). Iconic buildings continue to contribute to the image of the City as a business centre but can also draw visitors into the celebration of the 'world city'. In contrast with the emphasis on local distinctiveness elsewhere in the East End, at Canary Wharf and the Olympic City emerging at Stratford, city-wide image is more significant. In these projects the organic process of development of new quarters that we saw in Islington and Bankside is replaced by corporate management of controlled spaces.

5   New tourism (areas) in the 'New Berlin'

Johannes Novy and Sandra Huning

INTRODUCTION

Almost twenty years after the fall of the Berlin Wall, Berlin continues to be a city 'in the making'. Urban boosters' dreams after the City's reunification, that Berlin would become a 'global city' rivalling London, Paris, or at least Munich or Frankfurt have not materialised. *The Economist* (2006) put it pointedly that Germany's largest city and political capital 'resembles a glitzy shopping mall with lots of smart boutiques but no anchor tenant'. Heavyweight companies that had left Berlin after 1945, such as Siemens and Deutsche Bank have, despite the building and planning frenzy of the 1990s, seen no reason to return. Instead, Berlin struggles to this day to catch up with other economic centres of the poly-central German system, and is often regarded as a third-tier or gamma city in the global urban hierarchy (Krätke 2004 and Ward 2004).

However while Berlin's development since 1989 has in many regards been a story of broken dreams and unfulfilled expectations, there were also a number of unexpected successes. Berlin's rising status as an urban tourism destination is one of them. Tourism has, along with a few other industries, including global media, arts, music, and culture, as well as software and life sciences research, grown remarkably since the fall of the Wall, and has had a powerful impact on the City's profound transformation since then (Krajewski 2005). This impact is of course most evident in Berlin's central areas where most of the City's top tourist destinations are located, yet is also felt in areas that lack conventional attractions and were not planned – and until recently not marketed – as tourist zones. Former working-class districts at the inner-city's fringe like Kreuzberg, in former West Berlin, and Prenzlauer Berg, in former East Berlin, are cases in point (see Figure 5.1).

These areas share a reputation of being hubs of Berlin's alternative, bohemian, and creative scene, and are also associated with myriad myths and legends surrounding their history as places where people on the margins of East and West Berlin's urban society – voluntarily and involuntarily – came together. Developments in these areas lend support to the idea that urban tourism in western cities is increasingly characterised by patterns that

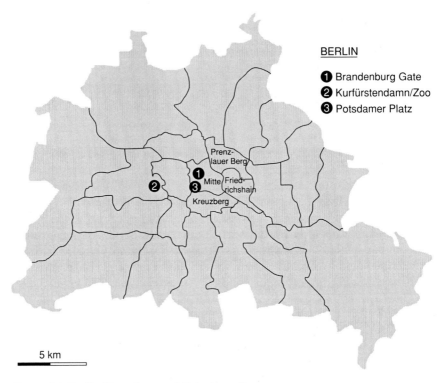

BERLIN

❶ Brandenburg Gate
❷ Kurfürstendamn/Zoo
❸ Potsdamer Platz

5 km

Figure 5.1 Berlin: Kreuzberg and Prinzlauer Berg
Source: Authors' own

are consistent with what has sometimes been described as 'new' or 'post' tourism and that urban tourism's spatial character is more complex than suggested by earlier scholarly accounts that emphasised the hermetic and regulated nature of leisure and consumption spaces (Lloyd 2002). Exemplifying a shift away from standardised mass tourism towards more individualised, differentiated touristic experiences (MacCannell 1999 and see also Rojek and Urry 1997), these neighbourhoods can be categorised as 'new tourism areas' (Maitland and Newman 2004) and illustrate how changing patterns of tourism consumption, *jointly* with post-industrial cities' changing consumption and production patterns and practices more generally, transforms the trajectories and character of inner-city neighbourhoods (see especially Lloyd 2002 and Maitland and Newman 2004).

What makes Berlin a fascinating case for the analysis of changing patterns of urban tourism, is that new tourism areas seem to have a somewhat different function in Germany's capital than in other world tourism cities where they typically play a subordinated role for cities' attractiveness as destinations. Lacking the lustre of London, or the beauty of Paris, Berlin as a visitor destination relies, as we will show, heavily on its reputation to be a city that is a dynamic, tolerant, diverse, experimental, and youthful place where anything goes and where trends are set (see Farias 2008 and Vivant 2007). While only attracting a certain, albeit continuously growing, share of Berlin tourists, hubs of the City's alternative, bohemian, and creative scene that possess a degree of distinctiveness and differ from the perceived 'placelessness' of planned destinations (Maitland and Newman 2004, p. 339), have come to play an important role for the 'destination' Berlin because they embody in public imagination many of the attributes that are associated with the City as a whole.

Embedded in a discussion of tourism's development in Berlin after the fall of the Wall, we will examine the role new tourism areas play for Berlin as a destination. Subsequently, an account of the way tourism has influenced the trajectory of Prenzlauer Berg will be provided and the role of the public sector interventions in influencing neighbourhood tourism discussed. In the second half of the chapter we will have a more detailed look at developments in Kreuzberg, West Berlin's former radical and multicultural centre. Particular attention will be paid to the socio-economic implications of tourism and leisure development in this disfavoured neighbourhood as well as to attempts by local actors to achieve sustainable, neighbourhood-based tourism.

BERLIN – WORLD TOURISM CITY?

Contrary to the enthusiastic rhetoric of urban marketing experts in the early 1990s, present-day Berlin's reality is disillusioning. Since the fall of the Wall, the City has lost two-thirds of its jobs in manufacturing, that today merely employs roughly 200,000 of the City's 3.4 million inhabitants (Statistisches Landesamt Berlin-Brandenburg 2007a). It suffers from the phasing-out of special federal FRG and GDR subsidies that had kept both parts of the City functioning during the cold war, and was, not least because of the political mismanagement and wastefulness that characterised the 1990s at the beginning of the twenty-first century effectively bankrupt. Unemployment and welfare dependency rates are high, socio-economic and spatial polarisation is rising, and the City's economy, despite recent signs of recovery, continues to underperform the German average (Mayer 2006a, Häussermann and Kapphan 2000 and Senatsverwaltung für Wirtschaft, Technologie und Frauen 2007). Yet, despite these socio-economic struggles, reunified Berlin clearly resumed its place as one of Europe's great capital, and tourist, cities and, in terms of its significance as a tourist destination, catapulted itself in only a few years into the top league of the global urban hierarchy.

As Germany's number one city destination for leisure and business tourism, the capital currently draws around 140 million visitors annually, placing the City third amongst other major European tourist hot spots behind London and Paris (see BTM 2007a). Of Berlin's approximately 140 million visitors in 2006, an estimated 126 million were day trippers (an increase of 20 percent compared to the year 2000); around seven million travellers spent the night with friends and relatives and more than seven million guests stayed in hotels, hostels and guesthouses (BTM 2007a) that accounted for a record-breaking 15.91 million overnight stays in 2006 alone, almost a million more than the City had initially expected to reach by 2010 (in comparison, 2004: 13.3 million, 2005: 14.6 million). According to unofficial estimates, both leisure and business tourism have thereby contributed to the City's increases in tourism activity with business tourism currently accounting for approximately 40 percent of Berlin's total number of visitors. International tourists currently account for approximately one-third of the City's overnight visitors and have significantly increased in recent years (38.8 percent in 2006 in comparison to 23.6 percent in 2001) in absolute numbers as well as relative to the share of domestic visitors (BTM 2007b). A representative survey of the Berlin Tourismus Marketing GmbH (BTM), that replaced Berlin's Public Transportation Office (Verkehrsamt) in 1994 as the City's official marketing authority, moreover reveals that Berlin is particularly attractive among young tourists: almost 50 percent of international visitors, and about a third of German visitors, are under thirty-five (BTM 2002). Having passed the target of 15 million overnight stays several years ahead of schedule, the local government and the City's tourist industry jointly decided to raise the benchmark for 2010 to 20 million overnight visitors, a number almost twice as high as 2000, and nearly three times higher compared to the years after the City's reunification (*Berliner Morgenpost* 2006). Given the increase in visitor numbers over the last decades, it comes as no surprise that tourism's economic relevance has also grown, both in absolute and relative terms. Tourism today contributes 7.5 percent of the City's aggregate income (2003: 5 percent), generates more than €1 billion in local taxes (2003: €703 million) and provides over 255,000 of the City's 1.5 million (fulltime) jobs (2003: 170.000) (BTM 2007b).

In a city with an unemployment rate between 17 and 20 percent, a notoriously tight fiscal budget, and relatively few growing industries, such numbers are of course highly appreciated (Statistisches Landesamt Berlin-Brandenburg 2007a). The promotion of Berlin as a tourist city has ranked high as a development and policy goal ever since reunification. For economic, as well as symbolic reasons, the City's élites soon after the fall of the Wall realised that tourism not only promised jobs and revenues, but could also help to redefine Berlin's identity and (re-)position the City in the German and international inter-urban competition (Häussermann and Colomb 2003). Embedded in broader efforts to turn Berlin into a post-industrial metropolis and a place of attraction for capital, creative industries, and talent, a 'tourism coalition' made up of private investors, the City government, urban marketing professionals, the media and other individual and collective actors in the course of the 1990s consequently engaged in various activities to promote the City as a destination and reorganise its urban landscape according to the needs of visitors and the tourist industry (see especially

Häussermann and Colomb 2003). Towards the end of the 1990s, as Berlin's falling competitiveness as a business destination, and the visitor economy's growing relevance for the City became increasingly evident, tourism gained even more importance in the City's governance and development arena. Wholeheartedly embracing creativity, culture and consumption as major production factors in boosting economic growth, the City and in particular the current Governing Mayor Klaus Wowereit who came into office in 2001, has aggressively pushed towards an expansion of the City's tourism trade. Among other things his governing coalition of Social Democrats (SPD) and Socialists ('Linkspartei', former PDS) launched several marketing initiatives, established a round table addressing issues related to tourism development and involved the City's key tourism stakeholders, and developed – in collaboration with the adjacent federal state Brandenburg – Berlin's first comprehensive tourism concept.

BERLIN'S TOURIST APPEAL

In 2006, the year Germany hosted the FIFA Soccer World Cup, on average 380,000 tourists visited Berlin every day of the year (BTM 2007a). For 2007, a year without any major event of international significance, experts expect this number to climb again. Statistics for the first six months of the year suggest that 2007 will be yet another record-breaking year for the City's tourism industry (BTM 2007c). Whereas Berlin clearly benefited from the historic events that unfolded in the late 1980s and early 1990s, like the fall of the Berlin Wall in 1989, Germany's reunification in 1990 and the subsequent buzz surrounding the return of the Capital function to the City, local experts often refer to the City's variety and affordability when trying to make sense of the City's tourist appeal. They emphasise that Berlin offers tourists a broad array of activities, experiences and facilities and does so at a relatively cheap price. Eating at a nice restaurant and drinking in bars is much less expensive than in Paris or London, the City's continuously expanding hotel industry is considered to be one of the most reasonably priced in Europe and getting to Berlin is cheap because of the decision of several low-budget airlines, such as Easyjet, to fly to and from Berlin. Centrally located and equipped with efficient access to different modes of transportation, Berlin, as a tourism destination, is rich in arts, heritage, entertainment, and recreational facilities with seventeen national, and more than 140 local museums, 300 galleries, 150 theatres, three opera houses, eight orchestras, extensive parks, lakes and waterways, several sport and entertainment arenas and hundreds of small concert venues, theatres and other off-mainstream locations. Additionally, Berlin is one of the world's most popular cities for conventions and congresses and a popular host destination for special and re-occurring international events of various kinds. These events not only draw millions of tourists by themselves, they also guarantee the City a priceless amount of publicity and media attention, the FIFA 2006 World Cup being a case in point. Berlin's affordability, its infrastructure and its broad array of amenities and attractions – its *tangible* characteristics and resources – are critical elements of present-day Berlin's tourist appeal.

Ultimately, however, tourists do not make their destination choices on the basis of a place's (unmediated) qualities. Instead, it is assumed that tourists 'act upon their image of a locality, rather than the "reality" of a destination' (Selby 2004, p. 66). Because of their importance to tourist destinations, place images have received increasing attention in the burgeoning tourism studies and related disciplines in recent years and various different definitions of what constitutes an image exist. In fact Gallarza *et al.*, cited in Selby (2004, p. 65), have argued that there are 'nearly as many definitions of place image as scholars contributing to its conceptualization'. Shields (1991, p. 61) for example defines place image as 'the various discrete meanings associated with real places or regions regardless of their character in reality', whereas Dichter (see Selby 2004, p. 67) states that 'an image is not just individual traits or qualities, but the total impression an entity makes on the minds of others'. While reliable data on tourists' individual or collective perception of, and associations with, Berlin as a destination are hard to come by, an analysis of the way Berlin is described in travel guides allows us to shed some light on the City's identity as a tourist destination (see Farias 2008). A central media of modern-day tourism communication, travel guide books along with other sorts of information play a critical role as 'markers' (MacCannell 1999) that reflect and affect the image of places as destinations. Hence an analysis of the way places are described within them provides valuable information about the conceptions and ideas that are associated with a destination (see especially Farias 2008).

Representations and Interpretations of Berlin as a Destination

Examining the depiction of Berlin in contemporary travel guide books, Farias (2008) argues that the identity of Berlin – or at least the identity attributed to Berlin in contemporary touristic guidebooks – is dominated by four central 'orderings' that he calls the 'haunted city', the 'always-becoming city', the 'green city on the water', and the 'Berliner Luft'. The notion of the 'haunted city' refers to Berlin's perception as an embodiment of recent world history, a 'palimpsest' of different times and histories packed with traces and memories of the Prussian Era, the Weimar Republic, Nazi Germany, the Cold War and so on, that allows visitors, unlike any other place in the world, to experience the twentieth century's ambiguous history (Huyssen 2003). Somewhat at odds with the City's image as a 'historical landscape', the notion of the 'always becoming city' refers to the widespread perception of Berlin as a place that is in permanent transition or, as commentator Karl Scheffler (1989[1910], p. 219) put it almost a century ago, as a place that is 'always in the process of becoming and never managing to be'. This includes both the City's past and present physical transformations as well as Berlin's image as a place of ongoing social and cultural change, experimentation and progress. The notion of Berlin as a 'green city on the water' highlights the natural landscape of Berlin and the City's surroundings with its expansive lakes, parks, and woods. Lastly, the much used phrase 'Berliner Luft' ('the air of Berlin') is introduced by Farias because it represents a widely known synonym for the City's free-spirited, culturally diverse, and tolerant 'climate' that is frequently cited as another defining feature of Berlin's appeal. Hence, what makes Berlin unique according to travel guides' portrayals are not just the City's history and its broad array of iconic new buildings, cultural attractions, parks, waterways and lakes. Instead, Farias' analysis illustrates that travel guides also regularly cite the City's supposingly ever-changing, tolerant, diverse and unconventional character as constitutive elements of its particularity and tourist appeal. This character is frequently discussed with reference to present-day Berlin's everyday culture and life in general and alternative, bohemian, creative and countercultural atmospheres and life-styles in particular. 'Off' scenes, in other words, as well as their physical mani-festations in the form of bars, cafés, design, shops, clubs, neighbourhoods and so on are covered prominently in travel guides as attractions on their own and a demonstration of Berlin's uniqueness (Vivant 2007). Local actors includ-ing the Berlin Tourismus GmbH (BTM) are keen to affirm this portrayal: 'Berlin inspires because of being unfinished, its spirit of optimism and the permanent change – so it has become a magnet for creativity and innova-tion right in the heart of Europe!' (BTM 2007d). This image gained further popularity in 2006 when the UNESCO appointed Berlin – as the first city in continental Europe – to the 'Creative Cities Network' under the frame-work of UNESCO's Global Alliance for Cultural Diversity and awarded Berlin the title 'City of Design' (UNESCO 2006). As an interface for, and inter-section between, a variety of cultures, lifestyles, and traditions, according to the UNESCO Berlin represents an attractive location for imaginative minds and 'a breeding ground for creative ideas' (UNESCO 2006). Irrespective of the fact that Berlin, as any other big city, has multiple and overlapping tourism roles and attracts all kinds of tourists for a variety of different reasons, this popular and *quasi*-official (self-) portrayal of Berlin has implications for the City's tourism trade: it affects the decision-making of potential tourists and hence the socio-demographic and lifestyle characteristics of Berlin visitors. It also influences Berlin visitors' (spatial) behaviour as well as the (spatial) character of tourism in the City.

BERLIN'S TOURISMSCAPE

As in most other big cities, tourism activity in Berlin concentrates in the City's most central districts. Berlin has two: Mitte, the nucleus of the City and centre of former East Berlin, and administrative, commercial, and governmental heartland of the 'New Berlin', and the 'City West', the centre of former West Berlin around the transport hub 'Zoologischer Garten'. While Berlin is a polycentric and spacious city, most of its main attractions like the Brandenburg Gate, Museumsinsel, Reichstag, Hackesche Höfe, Potsdamer Platz, or the shopping mile Kurfürstendamm as well as the majority of the City's roughly 600 hotels, hostels and guesthouses, are nonetheless located in these areas (Senatsverwaltung für Stadtentwicklung 2006). At the same time, tourism activity is not limited to the City's central districts. In comparison to most other 'world tourism cities', Berlin's landscape of urban tourism, or 'tourismscape' instead seems rather dispersed. Data on the spatial behaviour of Berlin tourists is scarce, yet evidence suggest that many Berlin visitors in fact do not limit their explorations to the City's central area(s). On the one hand this is to do with the City's complex and distinct topography and a relatively large number of conventional attractions and tourism-relevant resources – heritage sights; architectural icons, museums, hotels and so on – are in fact located beyond the City's core. On the other hand, it seems also to be a result of the particular touristic relevance of Berlin's everyday culture and life in general and its 'off' scenes and places in particular:

> To truly understand what makes Berlin tick, you must venture into its neighbourhoods. Watch Schöneberg yuppies stock up flowers and fresh veggies on Saturday's Winterfeldtplatz market. Listen to Turkish workers debate the latest soccer scores at a Kreuzberg café [. . .] Join students and counter-culturalists pondering their navels in bohemian Friedrichshain [. . .] test your stamina while clubbing with scenesters in trendy Mitte. Heck, go just about anywhere – with open eyes and heart – and you're pretty much guaranteed a fun time.
>
> (Schulte-Peevers 2002, p. 8)

Apart from the City's historic centre Mitte and centrally located areas in former West Berlin such as Charlottenburg, Wilmersdorf and Schöneberg that always possessed a well-equipped tourism infrastructure because of their central location but also because of their reputation as hubs of Berlin's alternative, bohemian, and creative scene. This is particularly the case in the multiethnic and culturally diverse Kreuzberg district and many neighbourhoods in former East Berlin such as Friedrichshain and Prenzlauer Berg that have become increasingly desirable as sites of tourism and leisure consumption. Data by the Department for Statistics (Statistisches Landesamt Berlin-Brandenburg 2007b) on tourism development in Prenzlauer Berg and Kreuzberg lend support to this observation as they show that both neighbourhoods experienced disproportional increases of accommodation businesses as well as overnight visitors and stays. In 2006, Kreuzberg welcomed 471,762 overnight guests and recorded 1,101,182 overnight stays in its twenty-seven hotels, hostels, and guesthouses (compared to 56,560 overnight guests (up 734 percent), 148,099 overnights stays (up 643 percent) and nine lodges in 1993). Meanwhile, Prenzlauer Berg recorded twenty lodges in 2006 and welcomed 248,315 overnight guests, 577,400 overnight stays (up from 18,306 guests (up 1,256 percent), 61,021 nights (up 846 percent), and three lodges in 1993). Additionally, experts estimate that the share of VFR tourists, as well as the number of guests staying in informal types of accommodation, is disproportionably high in both districts.

Prenzlauer Berg

Located to the north of Mitte, the former working-class district of Prenzlauer Berg fell into decay after World War II. During GDR times writers, artists and intellectuals established themselves in the district's dilapidated and cheap apartments and turned the area into a meeting place for those opposed to the East German régime (Roder and Tacke 2004). After 1989, illegal squats and secret unlicensed bars and clubs opened and turned the area into a focal point of the soon legendary 'Wild East' days of Berlin. Soon celebrated by national and international media as the 'funkiest part of town' (Bernt and Holm 2002), Prenzlauer Berg became a popular destination for the 'New Berlin's' young, bohemian and creative milieus. Small businesses, restaurants, galleries, theatres and shops moved in and the area underwent a revalorisation process that was boosted when large parts of the neighbourhood were designated as formal redevelopment areas (Sanierungsgebiet) worthy of preservation in the mid 1990s (Bernt 2003). Private developers refurbished the neighbourhood's old and previously derelict building stock, new residential and commercial real estate was built and the area became one of Berlin's most sought-after residential neighbourhoods among affluent, well-educated singles and double-earning couples, as well as a popular location for Berlin's cultural and creative industries. Towards the end of the 1990s gentrification accelerated and the area, although retaining much of its mythological reputation, became more bourgeois and decisively less bohemian.

Tourism played an important role in the district's transformation. Blending trendiness and heritage, as well as commerce and creative outsider elements, Prenzlauer Berg became increasingly attractive as a site of tourism and leisure consumption over the years, offering visitors both traces of the neighbourhood's 'legendary' past as well as the possibility to participate in, or simply gaze upon, post-reunification Berlin's alternative, bohemian/bourgeois and creative atmospheres and lifestyles. Today, tourism has taken hold in various parts of the neighbourhood, yet is particularly noticeable in the gentrification's initial footholds (Bernt and Holm 2002). This included places like Kollwitzplatz, a cobbled square surrounded by lavishly restored tenement buildings, nearby Kulturbrauerei, a former brewery that was converted into a culture and entertainment centre, or Kastanienallee (dubbed 'Castingallee'), a café-lined street where locals and tourists meet to 'see and be seen'. Also, as suggested earlier, a local tourist industry is emerging; numerous lodges have opened over the years for instance, among them many youth-oriented businesses like the 'Generator' that is, with over 850 beds, one of Europe's largest hostels.

Data on the characteristics, interests and activities of tourists who visit Prenzlauer Berg does not exist, however we argue that the development in the neighbourhood bears comparison with that is 'new tourism areas' in other world tourism cities (Maitland and Newman 2004). Areas like Kollwitzplatz, that serve as stages of the area's bohemian flair, and a few other destinations are frequented by a variety of different types of tourists with different backgrounds, yet most tourists seem to share characteristics with the new urban milieus that made the district their home and whose consumption preferences are stamped on its streets, both in terms of their demographic characteristics and their lifestyle and consumption preferences. Additionally, although a few tourist attractions (like the recently re-opened Rykestrasse Synagogue – Germany's largest Jewish House of Worship) exist, tourism seems not so much driven by particular sights and sites, but rather by many of the qualities that also attracted its increasingly affluent residential population and enterprises of the City's cultural and creative economy. This includes (but is not limited to) the area's social, cultural and physical environment, its atmosphere and small-scale amenities (cafés, galleries, shops, bars, and clubs) as well as the myths that surround it. Hence, Prenzlauer Berg's development challenges scholarly accounts that suggest a disjuncture between tourism patterns and practices and other land uses. Developments rather lend support to tourism research that stresses the increasing 'conviviality' (Maitland and Newman 2004) amongst different groups of city users in post-industrial cities' centrally located neighbourhoods, as well as gentrification research that highlights the significance of tourism – along with culture and leisure consumption more generally – as an important factor behind gentrifying neighbourhoods' transformation (see Zukin 1991 and 1995, Mele 2000 and Lloyd 2002).

Mutually reinforcing each other, Prenzlauer Berg's transformation into a tourism destination and preferred location for the 'new middle-classes' went hand in hand as tourism contributed to the neighbourhoods' revalorisation by underlining its aesthetic, cultural and symbolic qualities, and fuelling the proliferation of culture and consumption oriented amenities within it. That said, the district's transformation is, at the same time, by no means only an expression of contemporary patterns of culture and leisure consumption. Rather, it is also a result of government policy and planning. The public sector – although initially only marginally concerned with the area's potential as tourist destination – played a decisive role in its transformation by (co-)financing the upgrading of the neighbourhood's deficient housing stock, improving the attractiveness of its environment, and by investing into the expansion of its cultural infrastructure.

NEW TOURISM AND NEW URBAN POLICIES

When tourists began to venture into neighbourhoods like Prenzlauer Berg in growing numbers, tourism beyond the City's conventional sights and sites was hardly an issue in Berlin's governance and urban development arena. This has changed because actors concerned with tourism marketing and planning have recognised the importance of 'new tourism areas' and 'off-mainstream' places and scenes beyond the inner-city for Berlin as destinations. It is also the case that actors on the citywide, as well as on the district and neighbourhood level involving public sector actors as well as commercial and not-for-profit organisations, pay increasing attention to tourism – along with other strategies involving culture and consumption – to stimulate local economies, overcome socio-economic deficits, and further Berlin's position as a place of attraction for capital, creative industries, and talent. The State's government (Berlin is not merely a city, but one of three 'city-states', comparable with the other fifteen German *Länder*) launched a number of economic development programmes such as the 'Local Coalitions for the Economy and Labour' (Bezirkliche Bündnisse für Wirtschaft und Arbeit, BBWA) or the marketing initiative 'MittendrIn Berlin' to encourage the City's twelve borough councils – the second directly elected tier of government – to enhance their tourism marketing and invest in their touristic infrastructure. Moreover, a citywide signage system is being developed in order to facilitate way-finding throughout the City, and increased efforts by the creative and cultural industries unit in the Senate Department for Economy, Technology and Women can be observed to enhance the marketing of the City's spatially dispersed creative clusters to tourists.

Recognising the increased differentiation and segmentation of modern-day tourism, key players, in terms of city marketing, have also intensified their efforts to incorporate 'off the beaten track' places and scenes into their promotional activities, the above-mentioned Berlin Tourism Marketing GmbH (BTM) being a case in point. The public-private partnership's marketing is oriented towards numerous different target groups (travel groups, families, youths, gays and lesbians, cultural tourists, seniors, business travellers, congress participants etc.) and regularly promotes a broad range of attractions, activities, and amenities throughout the City. Although not formally concerned with tourism development, area-based neighbourhood regeneration programmes also devote resources to strengthen existing visitor economies and infuse new artistic and/or economic life to areas that had previously not been attractive as sites of tourism and leisure consumption.

Many of the City's local area management that were introduced within the scope of the federal/state 'Districts With Special Development Needs – the Socially Integrative City' programme (Soziale Stadt or Social City, for short) place an emphasis on the promotion of culture- and consumption-based development and try to exploit and boost neighbourhoods' cultural resources and entrepreneurial talent. In order to stimulate development in 'districts with special development needs', events, spectacles, and guided tours are organised, neighbourhoods' tangible and intangible assets advertised, and the development of artistic and creative scenes, from multimedia to galleries and theatres, are promoted. Such efforts are not uncontested. On the one hand, not all places have the potential to attract tourism activity and not all factors that need to be in place for areas to become attractive as sites of tourism and leisure consumption can be provided through policy and planning. Therefore, efforts to stimulate a local visitor and leisure economy do not always live up to their expectations. On the other hand, critics point out that this approach does not touch the conditions that lie at the heart of deprived neighbourhoods' problems, build on selective aesthetic and consumption preferences that are informed by middle-class values and lifestyles that are ultimately targeted at attracting more investment, as well as affluent consumers and residents, than at meeting the needs of long-time residents and businesses (Marcuse 2006, Mayer 2006a and Bockmeyer 2006). Against the background of a pronounced move towards social polarisation in Berlin in recent years (Mayer 2006a and Häussermann and Kapphan 2002), the implications of government policy in Berlin's deprived communities, and in particular their effects on different socio-economic group, clearly need to be explored in more depth. This is especially necessary in areas where urban renewal and regeneration policy as well as gentrification and increases in tourism activity are closely intertwined, as is so characteristic of Prenzlauer Berg's transition. In economic terms, however, this transition can be regarded as a success. Berlin as a whole benefits from Prenzlauer Berg's reputation, its booming creative industry (such as media, design, advertising etc.) as well as the 'lifestyle brand' the neighbourhood has become. At the same time, the spending power of those attracted to the neighbourhood boosts the local economy. Local actors in Prenzlauer Berg meanwhile are aware of tourism's significance. Co-financed through the European Union's EFRE fund, as well as federal and state funding, a tourism information centre was established in 2004 while the local business association 'Pro Prenzlauer Berg e.V.', that represents many of the businesses in the district's growing tourism industry, has worked on the marketing of the district since 1994. Intensified efforts to promote tourism development can also be observed in Kreuzberg in the City's western half that became (in)famous during the 1970s and 1980s as former West Germany's radical centre and focal point of Germany's Turkish diaspora. As illustrated in the discussion below, by the 1990s Kreuzberg had entered mainstream guidebook culture as a bohemian and multiethnic quarter of the 'New Berlin' yet, at the same time, faced a profound downward spiral of urban deprivation as the district's socio-economic fortunes declined.

THE CASE OF KREUZBERG

Located south of the City's historic centre, Kreuzberg as a district was formally established in 1920 as a union of parts of the former Luisenstadt, the Southern Friedrichstadt, and the Tempelhofer Vorstadt. Most of its housing stock stems from the late nineteenth and early twentieth century when industrialisation caused Berlin to grow exponentially. Far into the twentieth century, Kreuzberg was the most densely inhabited of Berlin's boroughs even in absolute numbers, with more than 400,000 residents and at times more than 60,000 residents per square kilometre. After World War II, large parts of the neighbourhood were destroyed and less than 60 percent of its tenement buildings were habitable (Diehl *et al.* 2002, p. 2). The City's division, and in particular the construction of the Berlin Wall in 1961, repositioned the borough as a rather isolated area of West Berlin. Because Kreuzberg was adjacent to the Berlin Wall, and much of the housing was dilapidated, rent was cheap. Turkish 'guestworkers' and other ethnic minorities, students, freaks and artists settled in the area's 'Mietskasernen' joining the disadvantaged inhabitants who could not afford to leave. The district became, in the words of Kil and Silver, an 'island of the foreign, the "Other", and the poor' (Kil and Silver 2006, p. 96) and, due to the high visibility of Turks, it became known as Berlin's 'Little Istanbul'. In the mid 1970s, when Kreuzberg had established itself as a byword for a unique mix of alternative lifestyles, multicultural scenes, young art and legendary nightlife, urban renewal and housing rehabilitation threatened to transform the area significantly. The south-eastern part of the district was most affected. Sometimes named 'SO 36' after the last two digits of its former postal code, the older housing in this area was cleared to make way for massive housing projects like the Neues Kottbusser Zentrum (NKZ) that still represents a nationwide symbol for the failures of the 'social democratic utopia' of the 1970s. At the same time, the area attracted more and more alternative and countercultural groups – draft-dodgers, punks, squatters and anarchists, left-wing activists etc. – who protested at the City's large-scale renewal plans and fiercely defended the neighbourhood's remaining Wilhelmian building stock. After years of conflict, the orientation of the Berlin state government shifted from new construction to renovation and preservation and from top-down to participatory planning (Kil and Silver 2006, p. 97). Squats were legalised, older buildings restored and Kreuzberg became an internationally recognised model project of sensitive urban renewal.

After Germany's reunification, the fall of the Wall put the district back in Berlin's geographic centre. Kreuzberg turned from a left-over niche at the edge of West Berlin into an inner-city district. Referring to its attractive location midway between the business centres of East and West Berlin, the local media proclaimed the upcoming transformation of the 'alternative mecca' into an 'exclusive yuppie district' (*Berliner Zeitung* in Lang 1998, p. 172). However, even though there were new investments and developments, gentrification did not occur to the extent that had been anticipated. Instead, Kreuzberg retained its image as a place of cultural pluralism and alternative lifestyles on the one hand and utmost marginalisation on the other. In the course of the 1990s, amidst economic restructuring, as well as stringent welfare cuts and budget restrictions by higher levels of government, the district's socio-economic deprivation became worse. The Berlin social structure atlas, that has been published regularly since the late 1990s, reveals that Kreuzberg has the highest level of social deprivation of all districts in the

City. Unemployment is 23 to 26 percent in some parts of the district with the proportion of under eighteen-year-olds frequently reaching double the City average. Problems related to the high concentration of immigrant residents in general, and the rising resignation of and tensions among youths with a migrant background in particular, who are especially isolated from economic and other opportunities, aggravate the perception that it is one of Berlin's most troubled areas. The public sector tries to address Kreuzberg's problems, for example through funds from various European Union programmes (e.g. ESF and EFRE), from different German funding programmes supporting the insertion of unemployed and welfare recipients into the labour market, programmes subsidising construction, as well as through the above-mentioned area-based federal/state initiative 'Socially Integrative City' ('Soziale Stadt'). In addition, a broad array of non-profit associations exist and provide services and engage in various activities to improve the neighbourhoods' socio-economic condition and counter disintegration and marginalisation processes (Bockmeyer 2006 and Mayer 2006b). Their efforts, as well as the funding provided through programmes like the 'Soziale Stadt' are, in light of Kreuzberg's structural disadvantages as well as the massive governmental cut backs of welfare provisions in recent years, widely considered insufficient to reverse the neighbourhood's decline.

Irrespective of the area's increasing socio-economic problems and its stigmatisation through local media outlets and politicians as a 'ghetto' and 'no-go area' (see Mayer 2006a), Kreuzberg never ceased to attract attention as one of Berlin's most vibrant neighbourhoods. Although facing competition from the 'newly' discovered bohemian neighbourhoods, such as Prenzlauer Berg, Kreuzberg's image as Berlin's 'multicultural melting pot' remained unchallenged. In fact, Soysal argues that Kreuzberg even gained importance as Berlin's 'stage for displaying diversity and multicultural flavour and colour' (Soysal 2006, p. 44) as actors, who, only a few years ago might have criticized immigrants' and non-traditional living arrangements, increasingly recognised that ethnic diversity, as well as diverse opportunities for cultural consumption, represent crucial assets as Berlin strives to become a cosmopolitan metropolis and position itself in what Harvey called the 'spatial division of consumption'. As 'new identifications have replaced the old idiosyncratic Berlin stories of a Divided City' and re-unified Berlin 'has been re-mapped in the image of a Hauptstadt of the unified Germany, Kulturstadt in a unified Europe, and Weltstadt in a cosmopolitan world', Soysal notes, Kreuzberg became the 'ceremonial ghetto for the metropolis' (Soysal 2006, p. 42).

The Kreuzberg tourist experience

Kreuzberg became a destination during the period when the City was still divided by the Wall and when tourists ventured into the area to participate in, or simply gaze upon, the neighbourhood's scenes, visit its galleries, pubs, clubs, ethnic markets and theatres (see Figure 5.2). Indeed, sightseeing

Figure 5.2 Berlin: Markets are part of Kreuzberg's appeal as a leisure destination
Source: Authors' own

coaches regularly drove by the area's squats and other manifestations of its 'otherness'. Its reputation as a world renowned hot spot of alternative, multicultural, and counter-cultural life, also guaranteed the neighbourhood coverage in travel guides and other travel-relevant media. This was particularly evident in guide books oriented towards readers with a more 'alternative' background that informed readers about its role as a headquarters of minorities and non-conformist groups, its inherent social problems and dilapidated housing conditions, as well as particular destinations such as pubs, galleries, and theatres (Lang 1998, 139). After the Wall came down in 1989 the district lost its insular character as well as parts of the scenes that had come to define its image. At the same time, developments in the previously neglected, but now geographically attractive, no-man's land, between Mitte and Kreuzberg as well as along the River Spree (including several hotels and Daniel Liebeskind's internationally acclaimed Jewish Museum which opened in 2001), led to new visitor streams but at the same time furthered concerns about a gradual dilution of the neighbourhood's distinctiveness. Elsewhere, developments were less cataclysmic than at the district's fringes and Kreuzberg remained 'in the (self-) portrayal of the City as the locus of hip and diversity' (Soysal 2006, p. 43). The widely available youth guide, *Berlin for Young People*, for instance, describes the district as a 'multicultural mix of peoples, Turks [living] along with students, "alternatives", punks, and perfectly normal Berlin families', and 'off-movie houses and theatres, wonderfully dingy bars, affordable restaurants and second-hand shops' (Herden Studienreisen Berlin 2006, p. 43). Today, tourism demand appears highly diverse. Quests for knowledge and understanding of migrant culture and Kreuzberg's past, as West Berlin's radical centre, co-exist with more mundane quests for a fun night out or a shopping spree at the district's broad array of bazaars and specialised stores.

Local events, like the annual Carnival of Cultures (Karneval der Kulturen), have furthered the district's profile as a destination. Steadily growing each year, both in terms of participants and spectators, the carnival gathers together community and minority groups, cultural initiatives, youth centres and other non-profit associations for a four-day festival to celebrate the diversity of Berlin (Soysal 2006, p. 43). Currently in its thirteenth year of its existence, Germany's biggest multicultural festival has become a fixture in Berlin's packed event calendar and is celebrated by locals and visitors (ironically described by the carnival's organisers as 'Problemkiezbewohner' and 'No-go-area-goers'), alike. Different opinions exist about the carnival and the 'staging of diversity' it involves (see Soysal 2006). Some reflect more general concerns about the distortion and banalisation of local ethnicity, culture, and everyday life as a consequence of tourism and leisure consumption. Others reflect recollection by residents of the time before the fall of the Berlin Wall when tourist buses drove through the district to let tourists have a glance at the neighbourhood's squatted buildings and its counter-cultural atmosphere. Ultimately, however, conflicts between locals and visitors are rarely heard of. Instead, tourist activity to this day seems rather well integrated into the local social fabric and built environment. Few 'tourist traps' exist and tourists patronise the same shops, bars, cafés and cultural attractions as local residents as well as visitors from other parts of the City without attracting much attention.

Why tourists keep a relative low profile when visiting Kreuzberg is difficult to assess, particularly given the insufficient data on the number and type of tourists that visit the area. We assume however that Kreuzberg's transitory character and diverse residential composition, as well as similarities between tourists and parts of Kreuzberg's residential population in terms of their demographic background, lifestyle and behaviour, are important parts of the explanation. One consequence of tourists' low profile in the neighbourhood has been that their economic contribution in the past has rarely been accounted for and they are frequently not even recognised as tourists since they overwhelmingly travel on their own and blend relatively well into the neighbourhood's life.

Tourism Policy in Kreuzberg

In many ways present-day Kreuzberg epitomises a contested phrase once used by Berlin's Mayor Klaus Wowereit to describe Berlin as a whole: 'the district is poor but sexy'. In other words, the district, and its residents, possess little capital in strictly economic terms, but the area is rich in aesthetic, cultural, and symbolic resources. Until recently, relatively little attention was paid on the local level to the convertibility of the neighbourhood's assets in economic capital – for example their utility as a source of income and job generation. Local actors in the past refrained from instrumentalising endogenous resources through marketing or similar activities to generate economic impulses. Ever since the upheavals against the City's urban renewal plans of the 1970s, development politics in Kreuzberg have been strongly influenced by a local élite made up of (former) community activists, left-wing academics and other actors that were involved in the battles of the 1970s and 1980s. Many of these actors are firmly established in Kreuzberg's governance and development arena – for example as members of the Green Party that regularly wins the largest shares of votes in most of the district's electoral districts. Others are organised in local development corporations and other organisations that influence local development. This élite advocates 'neighbourhood' interests such as community services, education, housing and participatory planning, and has in the past frequently played a pivotal role in protecting the area and its residents from developments that were deemed exploitive or inappropriate by the milieus they represent. In addition, a new generation of community groups, activists, and movements exists. Their influence becomes particularly visible during May Day, the annual – and until recently frequently violent – Labour Day celebration of leftwing groups that takes place on Kreuzberg's streets and was also exemplified by recent clashes between the burger chain McDonalds and a local initiative under the name McWiderstand ('McResistance') that fought the company's plans to open its first restaurant in Kreuzberg.

The group's ultimately unsuccessful fight against McDonalds (the burger chain opened its restaurants in September 2007) rendered visible once again Kreuzberg's protest culture that characterised it in the 1970s and 1980s. Yet despite Kreuzberg's reputation as a stronghold of leftist activism, sentiment and widespread scepticism against growth- or market-oriented development, local development politics in Kreuzberg in reality are of course not all that different from other parts of the City: on the one hand because the neighbourhood's development is – and of course always was – subject to developments beyond its borders such as politics and policies of higher levels of government. Thus, local actors' influence is limited and has as a matter of fact decreased over time after Kreuzberg as a municipal borough was merged with adjacent Friedrichshain in the course of an administrative reform that reduced the number of boroughs from twenty-three to twelve in 2001. At the same time, the broader changes in urban policy making and development that have characterised recent decades have not left Kreuzberg untouched. The rise to prominence of culture and consumption as engines of urban economic growth, and a move away from a paternalistic, welfare-oriented model of urban development (with direct state involvement) to a more privatised model emphasising growth, self-help, entrepreneurialism and competitiveness are cases in point. Against this background, and in view of the district's socio-economic troubles, budget restrictions and the decline of social welfare systems, local actors in Kreuzberg have in recent years intensified their efforts to deploy Kreuzberg's endogenous resources and its residents' entrepreneurial talent as means of neighbourhood development (see Figure 5.3). In line with developments in other areas, a growing number of actors in this context also turned to tourism development in order to generate jobs and revenues for local residents and businesses and counter the neighbourhood's deprivation.

From 2003 to 2006, the Friedrichshain-Kreuzberg's Mayor for Economics supported an initiative titled 'District tourism as a means of income generation in Friedrichshain-Kreuzberg' that aimed to promote tourism in the borough in order to spur the local economy and create job and revenue opportunities for local residents. Co-financed by the European Union, a local tourism marketing organisation – Multi-Kult-Tour e.V. – was established to reach out to members of the tourism industry and related sectors, community development organisations and social service providers, as well as public sector actors. In order to foster communication and cooperation, Multi-Kult-Tour e.V organised regular round table meetings in which the members of the

Figure 5.3 Berlin: Streetlife on one of So36's main thoroughfares Oranienstraße
Source: Authors' own

association discussed issues related to tourism development and ways to stimulate the sector. Moreover, tourist maps and brochures and a website (www.multi-kult-tour.de) were developed, small and large-scale events organised and previously unemployed residents were trained in skills like public relations, marketing, and fundraising to prepare them for jobs with decent earning potential. After the funding period through the public sector ended, the non-profit association activities were scaled back and it remains to be seen whether the working relationships and networks that were established during the funding period among tourism-relevant actors are sustainable and will lead to tangible results. More bottom-up oriented, local grassroots organisations are also increasingly building on tourism development. The Regenbogenfabrik ('Rainbow Factory'), a community centre that emerged out of the squatter movement of the 1980s, is a case in point. Based on a non-profit, local economic approach, the Regenbogenfabrik has, in the course of the last few years, expanded its range of activities and developed several tourism-related services to provide employment opportunities for previously unemployed people with different qualification levels and from different ethnic backgrounds or origins. Supported through public funds like the European Union's EQUAL programme, formerly unemployed people currently work at the Regenbogenfabrik's hostel and bike-rental shop, offer guided walking and bicycle tours, and moreover receive training in tourism-related subjects like foreign languages, tourist guiding, marketing, computer skills and book-keeping.

Job-training and orientation is also at the core of the tourism initiative 'Ich bin ein Berliner' ('I am a Berliner') that was developed by the non-profit association FIPP e.V. and offers different guided tours by local youths. Youths present Kreuzberg and other parts of Berlin from their individual everyday perspective and by doing so provide an alternative to the commercial tour enterprises that operate in the district. The 'X-Berg Tag' tour, one of the walks the initiative offers, includes visits to a mosque, a tea house, an ethnic restaurant, local shops and the Kreuzberg Museum that operates under the auspices of the borough's government and whose permanent exhibition informs visitors about the social history of the neighbourhood. In this project, as well as several others, tourism not only assumes a socio-economic, but also an informal educational, function. These, and several other initiatives, however small and unprofessional, exemplify a growing awareness of local actors in Kreuzberg concerning their neighbourhood's attractiveness as a tourist destination as well as tourism's potential for local development purposes.

Kreuzberg at the Crossroads

Different discourses shape popular thinking about the Kreuzberg of today. One perspective stresses the neighbourhood's socio-economic deprivation and characterises Kreuzberg as a symbol for everything that's wrong about

(West) Germany's immigration history. According to this perspective, Kreuzberg is on its way to become a 'ghetto' (or has already become one) where Turkish people and other minorities are being isolated and isolate themselves, where social problems such as unemployment, poverty, violence, and vandalism dominate, and where outsiders are less and less seen. In contrast, a second perspective depicts Kreuzberg as Germany's only truly multicultural community and a casual place of new lifestyles, multicultural creativity, and interethnic tolerance. Kreuzberg's ills are not necessarily neglected, yet its qualities are stressed. Sometimes, as for example in the above mentioned guide 'Berlin for young people', even Kreuzberg's problems are turned into an asset: 'Kreuzberg is the poorest district – so it's gotta be super sexy!' (Herden Studienreisen Berlin 2006, p. 129). According to this perspective, Kreuberg is alive and well and about to resurge as creative scenes that partially left in the early 1990s, in favour of the City's eastern districts, are rediscovering the neighbourhood and migrant communities are infusing new energy into the area's cultural scene. As is usually the case with such disparate views, Kreuzberg's reality probably lies somewhere in between these accounts. Clearly, the challenges Kreuzberg faces are considerable and the problems that resulted from the disappearance of industrial work, the decline of social welfare systems, failed integration and concentrations of ethnic poverty cannot be ignored.

By spending money in local shops, cafés and restaurants, by providing revenues for its (socio-) cultural scene, enhancing the recognition and value of Kreuzberg's cultural and ethnic resources inside and outside of the community and, last but not least, by challenging the stigmatisation of Kreuzberg as a 'no-go area', tourism has helped to lessen processes of disintegration and deprivation. Efforts by local grassroots like the Regenbogenfabrik illustrate how tourism development is increasingly integrated into 'bottom-up' development approaches under local control. These projects provide jobs and job training opportunities for those local residents that need it most and improve the neighbourhood's tourism-relevant infrastructure. To be sure one should not exaggerate the impact of these efforts however, as most of them are small in scope and their effectiveness is unclear. Additionally it is worth remembering in this context that districts and neighbourhoods are not collective actors which is why an examination of tourism and leisure developments' effects must look closely at the distribution of costs and benefits among different groups and actors. As of now, the Turkish community and other immigrant groups – who are critical sources for Kreuzberg's diversity and vitality – are only insufficiently integrated in attempts to expand the local visitor and leisure economy and we do not know to what extent these groups benefit from tourism and how they feel about the marketing of their community to high-spending visitors.

Generally, the lack of a strategic, integrated policy framework concerned with tourism 'off the beaten track' for the district – but also for the City as a whole – is noticeable. This is in contrast with developments in some other cities that are using strategic and comprehensive planning to stimulate tourism development in their multicultural and bohemian districts (see, for example, Shaw *et al.* 2004). The potential to tap into the growing market of tourists with a migrant background, who already seem to constitute a sizeable share of consumers in Kreuzberg's specialised stores and restaurants, has so far been ignored. That said, the lack of tourism planning and marketing, and the low profile tourism has maintained over the years, also seems to represent one of the reasons for the area's appeal and a further increase of tourism activity might give rise to many of the adverse effects that have led many commentators to be critical of tourism development in marginalised neighbourhoods. Several scholarly accounts can be found that stress that tourism revalorises marginalised neighbourhoods but does so at the expense of long-time residents and businesses. This is because most of the benefits are reaped by local élites and community outsiders while commercial and residential gentrification, rising living costs and other adverse effects put additional burdens on those community members that are most in need (see Zukin 1995, Gotham 2005, Dávila 2004 and, for a discussion, Shaw *et al.* 2004). These and other concerns, such as those about a homogenisation of the built environment and a distortion and dilution of local culture and identity, have to be taken seriously. What we find in Kreuzberg however is that tourism – frequently invisible – benefits the neighbourhood in manifold ways, amongst other things by contributing to a greater recognition of the area's qualities and resources and by helping to keep up a positive image of its living settings and culture. Whether this image will lead to developments like those in Prenzlauer Berg, – increasing property values and gentrification – remains to be seen. It ultimately depends not simply on tourism, but on the further development of Berlin's economy and property market, the political will of relevant actors to resist the current doctrine of market-oriented growth and state deregulation, and their inclination to instead commit to socially equitable development (see Huning and Novy 2006).

CONCLUSION

Reflecting widespread changes with regard to contemporary patterns of urban development as well as travel, tourism has become an integral component of everyday life in areas like Prenzlauer Berg or Kreuzberg and has played a substantial part in their development in recent years. Along with other parts of the City, like Friedrichshain, that has undergone similar developments and also attracts attention as a centre of Berlin's young, creative scenes, these areas not only illustrate the changing face of leisure and tourism consumption in contemporary cities, but have come to play an important role for the 'New Berlin' as a 'world tourism city'. Not only as supplements to the City's conventional attractions but also as embodiments of various attributes that have become trademarks of Berlin's image as a destination, these neighbourhoods make a significant contribution to the City's tourism trade. Tourism, evidence suggests, meanwhile contributed to the visible reconfiguration of many of Berlin's centrally located former working-class neighbourhoods, amongst other things by underlining their aesthetic, cultural and symbolic qualities, and providing revenues for consumption-based and creative industries within them.

Not all visitors to Berlin venture into the City's new tourism areas however, and more research is necessary to better understand who these tourists are, how they behave and the reasons for their visit. Based on what we know, few absolute statements can be made. What we can say with certainty is that most tourists who spend time in areas like Prenzlauer Berg or Kreuzberg do not fit into the negative polemics that have shaped much of the thinking about tourists in the past. 'Typical tourists' in Berlin's new tourism areas do not wear Hawaii-shirts, take lots of pictures and travel in large groups. While some of them might come to gaze upon 'the other' (be it immigrants, bohemians, punks, or scenesters), most lend support to scholars who advocate a more interactive and performative model of what tourists do and how they behave, blend into the local social fabric and patterns of everyday life, and appear to share many of the demographic qualities and consumption and lifestyle preferences of workers and residents in the neighbourhoods. Of the neighbourhood's *middle-class* workers and residents that is. In districts like Kreuzberg middle-class inhabitants are a minority and even in Prenzlauer Berg a large number of low-income residents defy the gentrification waves that were set in motion in the course of the 1990s (see Bernt and Holm 2002). Hence, the observable conviviality among the more privileged city users in 'new tourism areas' must not distract from the situation of poor residents and other disadvantaged population groups and the effect a rise in tourism and leisure has on them.

Clearly, more research is necessary to examine the socio-economic implications of tourism and leisure development in Berlin's centrally located neighbourhoods. This is particularly relevant given the rise of tourism and leisure development in Berlin since the early 1990s, and the City's parallel transformation from a relatively compact, socially integrated city to a fragmented cityscape with more pronounced social and spatial inequalities. Meanwhile, the effects of efforts to utilise an area's tourist potential as a means of local regeneration are meanwhile open to question. Although mostly small and not comprehensively planned, some of the approaches employed in Kreuzberg appear to make a serious effort to develop tourism to benefit local residents and businesses, yet it is simply too early to accurately assess whether they will succeed to better exploit the potentials tourism brings about while avoiding potential pitfalls. Additionally, uncertainties remain about the sustainability of some of these efforts in the face of the visible increase in competition between city districts within Berlin as well as by comparison with other types of investment and potential uses of resources.

6 Sydney

Beyond iconicity

Bruce Hayllar and Tony Griffin

INTRODUCTION

Sydney is a quintessential post-modern city incorporating modernist architecture and landscapes wrapped around the post-modern sensibilities of its people and economy. It is the financial capital of Australia and its only 'world city'. At its core is Sydney Harbour and its associated icons, the Harbour Bridge and Opera House. Together, these form the City's international persona and are amongst the most recognisable of all urban sights. The relatively mundane Central Business District (CBD), dominated by international-style, high-rise office towers, is lifted aesthetically by its juxtaposition to the harbour, that surrounds it on three sides. With the harbour as its centre point, it is little wonder that Sydney has become a major international tourism destination.

However, Sydney's emergence as a tourism destination is relatively recent when compared to other world cities discussed in this book, such as London, Paris and New York. It is only since the mid-1980s that tourism has become significant for the City and a major influence on its social and physical fabric. In contemporary times it is hard to imagine Sydney without tourists and the overt signs of commercial tourism enterprises, particularly around the harbour and central-city areas. Sightseeing cruises, pleasure craft of all shapes and sizes, and the ubiquitous jet boats for the more thrill-seeking tourists, dominate the harbour. Tourism and leisure activities now occupy an almost unbroken stretch of foreshore land extending nearly ten kilometres from Woolloomooloo to Pyrmont. Within this stretch of foreshore are many of Sydney's principal tourist attractions, including the Royal Botanic Gardens, Sydney Opera House, Circular Quay, The Rocks, the Harbour Bridge, Walsh Bay arts precinct, King Street Wharf, Darling Harbour and the 'Star City' Casino (see Figure 6.1).

Significantly, much of this waterfront land and its immediate environs are under the ownership or management of a state government, the Sydney Harbour Foreshore Authority (SHFA), reflecting the importance that the New South Wales (NSW) government places on this area and its tourism-related functions. Indeed, the way in which SHFA has evolved over time from

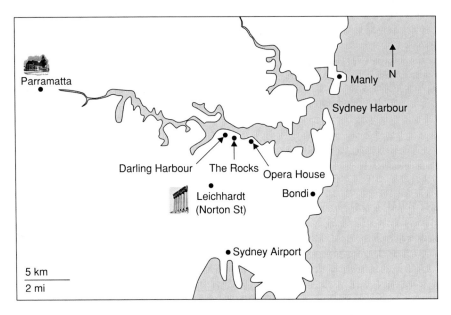

Figure 6.1 Sydney study precincts – locations
Source: Authors' own.

its antecedent authorities is symbolic of the growing importance of tourism to the Sydney and NSW economies over the past few decades.

TOURISM IN SYDNEY: AN HISTORICAL PERSPECTIVE

In spite of tourism's increasing prominence and profile, its development in Sydney has rarely been guided by conscious and deliberate policy. Major public policy initiatives have certainly influenced tourism, but often tourism has been a by-product rather than a strategic objective central to these initiatives, particularly prior to the mid-1980s. This section of the chapter reviews the development of tourism in Sydney decade by decade since the 1960s and major factors and events influencing the development of tourism within the City are highlighted, including the role of government policy.

1960s and 1970s: The 'Accidental' Tourist Destination

In the 1960s, Australia was fairly inward-looking and, as a city, Sydney reflected this. To the extent that the nation gazed outward at all, it did so primarily in two directions: 'Mother England', the progenitor of the national culture and home of Australia's head of state, and the USA, the country's most important military ally. To Britain, Australia owed the debt of its identity and the economic benefits that membership of the British Commonwealth afforded it and to which remained stoically British and loyal. To the USA, Australia owed its survival as a sovereign nation during World War II and its ongoing security, given its isolated position at the base of Asia. South-east Asia in particular had experienced a turbulent period with the end of European colonialism and the emergent threat of communism. In Vietnam, America had responded militarily to these events, with Australia's enthusiastic and grateful support. While Britain represented the past, the USA represented the future, politically and economically, particularly with Britain showing signs of interest in joining the European Economic Community and in the process abandoning the favourable trade terms it hitherto offered Commonwealth nations.

In the early 1960s Sydney strongly reflected its British heritage in both form and feel, although the American influence was showing signs of taking root. Tourism was primarily domestic and barely noticeable. Sydney's beaches, particularly Manly and Bondi, attracted summer holidaymakers and were popular day-trip destinations for 'Sydneysiders'. Much of the central-city consisted of sturdy Victorian sandstone buildings and was fundamentally concerned with the serious business of property, finance and high order retailing. Entertainment was confined to the corner pubs, that still closed their doors at 6pm, and a concentration of cinemas and theatres at the southern end of the CBD. The occasional steakhouse or smoke-filled jazz club could be found nestled in one of the City's more intimate laneways, but most such activities were located a couple of kilometres away in the Kings Cross area.

Kings Cross, at this stage, had developed as something of an alternative, bohemian quarter. The home of many artists and students, it had a cosmopolitan café culture and nightlife that stood in strong contrast to the rest of the City. Kings Cross was effectively the City's first truly urban tourism precinct, based largely on its concentration of nightlife and entertainment activities and its somewhat 'racy' reputation. The business of prostitution and strip clubs received a particular boost during the Vietnam War years, when thousands of American servicemen would regularly visit the City on leave. Sydney's first major hotels catering to leisure travellers were also constructed in this area. Famously, the Beatles had waved from the balcony of the *Kings Cross Sheraton* on the first morning of their 1964 visit to Australia, while the nearby *Chevron Hilton's Silver Spade Room* played host to most of the major international performing artists for more mature audiences. The CBD, on the other hand, contained only the austere *Menzies Hotel* and a handful of other establishments for the business traveller.

Around the harbour, some significant changes were afoot. On Bennelong Point, the old tram sheds had been demolished, no longer needed because of the decision to dismantle the tramways in 1961 in order to free up the City's streets for the motor car (see Nixon 2008). For more than a decade Sydneysiders watched, sometimes eagerly and sometimes sceptically, as the distinctive form of the Opera House gradually took shape on this site. The brainchild of J.J. Cahill, a socialist NSW Premier who wanted to make high culture accessible to the masses, the Opera House was reflective of a more progressive Australia that was breaking away from its conservative roots.

Starting in the late 1950s around Circular Quay, the area juxtaposing the Opera House site, Sydney's first 'skyscrapers' were being built. This trend gathered pace throughout the 1960s and early 1970s as sleek office towers began to replace much of the Victorian fabric of the City, spurred on by a burgeoning finance and property sector and significant levels of foreign investment. Such development was seen as indicative of Sydney's trans-formation into a modern, progressive, international city, with pretensions to be the financial capital of the South Pacific. It was in this climate that the future of an area that was to become one of Sydney's most important tourism precincts took a dramatic change of direction.

Located on the harbour, at the northern end of the CBD and in the shadow of the Harbour Bridge, was an area known as The Rocks (see Figure 6.2). In the 1960s The Rocks was widely regarded as a crime-ridden slum of decrepit and derelict buildings. It was perceived as ripe for redevelopment, in spite of it being the site of the original European settlement in Sydney and con-taining the oldest extant buildings of colonial origin in Australia. Fortuitously for the development-minded NSW government of the late 1960s, most of the area was under public ownership, having been resumed by the government in the early 1900s in order to stem an outbreak of bubonic plague! A new

Figure 6.2 Sydney: The Rocks
Source: Authors' own.

statutory authority, the Sydney Cove Redevelopment Authority (SCRA) was created by an Act of the NSW Parliament in 1970 and given control over The Rocks. SCRA's charter was to plan for and carry out the wholesale redevelopment of the area and maximize the financial return to the government from doing so. Ownership of the land was to remain in public hands, with all development proceeding under leasehold arrangements. As landlord, SCRA possessed virtually absolute power over the site and was accountable only to its Minister. The creation of such a powerful statutory authority was intended to expedite the process of redevelopment, to ensure that it proceeded quickly and smoothly. SCRA's plan was basically to raze the existing built fabric and extend the CBD northwards through the construction of high-rise office towers. That The Rocks of today does not reflect this plan is largely an accident of circumstances converging to frustrate the government's original intent.

In the early 1970s, the demolition crews moved into The Rocks only to be confronted by an alliance of residential tenants, some of whose families had lived in the area for generations, and a militant trade union, the Builders Labourers Federation (BLF). The BLF had, on a number of previous occasions in Sydney, shown its willingness to support residents against development interests and had placed a 'green ban' on The Rocks that effectively barred any union member from working on the demolition or construction (Roddewig 1978). Green bans had evolved partly because the state planning legislation of the day afforded local communities virtually no rights to object to, or otherwise influence, planning and development decisions. The green ban succeeded in delaying the redevelopment until a collapse in the commercial property market in 1973/1974 raised questions about the viability of the original plan. At a time of massive oversupply in city office space it would not have served the government's financial objectives for The Rocks to release further space onto the market. By the time the office market had recovered, a new state government, more sympathetic to conservation of heritage,[1] was in office. The Rocks was now recognised for its important heritage values, but SCRA still needed to generate a financial return from the government's substantial investment. The plan then became one based on restoration of as much of the built fabric as possible. Demolition would only occur where buildings could not be safely restored, and new constructions had to fit into the scale and character of the historic built fabric. This in turn limited the land use options that were available to occupants of buildings in the area. Almost by default, tourism was now seen as the principal commercial activity of the area, and restaurants, cafés, and specialty retailing became the major tenants of the renovated buildings. On the southern fringe of the precinct, adjacent to the CBD, a number of sites were earmarked for future major hotel developments. SCRA began to promote the area as an historic tourism precinct, the 'Birthplace of Australia', and its commercial success was now contingent on attracting tourists who would spend money in the tenant businesses. Sydney now had its first mainstream urban tourism precinct.

1980s: The World Discovers Sydney, and Sydney Discovers Tourism

In spite of the emergence of The Rocks as a tourism precinct, by the early 1980s Sydney was still an international tourism backwater. In 1985, the City boasted only eight four- and five-star hotels, and even that limited supply had been boosted by the opening of the *Regent* and *Inter-Continental* hotels in 1983 and 1985 respectively (Griffin 1989). All that, however, was about to change. In 1985, Australia received some one million international visitors. By 1988, that number had more than doubled as international arrivals grew at annual rates in excess of 25 percent throughout the middle part of the decade. As the major international gateway, Sydney was receiving a substantial proportion of that growth and featured in the standard itinerary for many package tourists – the classic 'Sydney, the Reef and the Rock'.[2] On the back of that growth, and its anticipated continuation, the supply of up-market hotel rooms in the CBD alone was set to more than triple from 1988 to 1993, with nineteen new hotels containing nearly 8,000 rooms either under construction or in the planning pipeline at that time (Griffin 1989). More budget accommodation was planned around the fringes of the City centre.

The 1980s also saw the development of Sydney's most substantial commitment to tourism – Darling Harbour (see Figure 6.3). Located to the

Figure 6.3 Sydney: Darling Harbour
Source: Authors' own.

immediate west of the CBD, Darling Harbour had been a blighted area of redundant dockland, railway goods yards and warehouses since the 1970s. In the early 1980s, however, plans were announced to redevelop the area into the centrepiece of Sydney's celebrations of the bicentennial of European settlement. Thus, in 1984, the NSW government created the Darling Harbour Authority (DHA) to implement this plan. Like SCRA in The Rocks, DHA was given extraordinary powers as a combined landlord, developer and planning authority for the site. To circumvent the possibility that its redevelopment could be delayed by public objections, Darling Harbour was exempted from the normal provisions of the State's environmental planning legislation. Statutory planning obligations such as placing plans on exhibition, allowing for public submissions, objections and rights of appeal and environmental and social impact assessment did not have to be included in Darling Harbour's planning and development processes as a matter of deliberate policy. Moreover, the Sydney City Council, the local government authority for the area, was effectively excluded from any part of the planning process. All this was intended to ensure that the new Darling Harbour was ready in time for the major bicentennial celebrations in January 1988 and it was a feature of its development that drew much criticism (Hall 1998).

Darling Harbour was certainly intentionally created as a tourism precinct, although it also represented a significant exercise in the urban renewal of a blighted area. Its size and strategic location on the harbour and adjacent to the CBD provided an opportunity to develop a range of major facilities that the City was perceived to lack. Hence, the government invested heavily in facilities such as the Sydney Convention and Exhibition Centre and National Maritime Museum, to enhance the City's capacity to attract visitors. The private sector also contributed to the development, and at the time of its opening, Darling Harbour contained a number of other major attractions and facilities, such as the Sydney Aquarium and Harbourside shopping centre that incorporated numerous cafés, bars and restaurants. The area also featured significant pockets of open space, including spaces for outdoor performances and events. Many of the aforementioned new hotels were constructed in and around this new tourism precinct.

Darling Harbour, however, was not an immediate success. In its initial year of operation it drew some 13 million visitors, less than had been anticipated, and by 1995 this had risen to only 13.7 million (Darling Harbour Authority 1996, cited in Hall 1998), with the majority being 'Sydneysiders'. Some of its key elements, notably the Harbourside shopping centre, have proved to be persistently problematic from a commercial point of view, with numerous changes in the management, configuration and tenant mix over the years. A few of the originally planned developments did not eventuate, in one case due to the financial collapse of the private developer. Rather than being a *fait accompli*, by the end of the 1980s the precinct was effectively developed in a number of stages over the next two decades and it was well into the 1990s before there was a sense of completion about it.

1990s: And the Winner is . . . Sydney!

For tourism in Sydney, the 1990s was a decade of extremes. International tourism had temporarily plateaued after the euphoria of the 1988 Bicentennial year. While growth resumed in the early 1990s, it was much more modest than the heady days of the mid-1980s. Many of the new hotels, whose development had been stimulated by this growth, opened in conditions where supply had far outstripped demand. As a consequence, some were in receivership even before they had welcomed their first guest. Financiers became hotel owners and numerous new properties were put on the market for substantially less than their development costs as the bankers sought to recoup their losses from tourism investors and developers unable to service their loans (Griffin and Darcy 1997). After the boom of the 1980s, Sydney was experiencing its first tourism development bust. The hard times for Sydney tourism were, however, short-lived.

Early on the morning of 24 September 1993, tens of thousands of Sydney-siders had gathered around Circular Quay and the Opera House. The crowd fell silent for a moment as the President of the International Olympic Committee, Juan Antonio Samaranch, appeared on the massive TV screens dotted around the area, only to erupt when he announced that Sydney had won the right to host the 2000 Olympic Games. Aside from the local populace's outpouring of delight at this news, the outlook for tourism in Sydney once again looked very positive.

Securing the right to host the Olympics represented a major opportunity for Sydney, from both tourism and urban development perspectives. However, the decision to bid for the Games was not strongly motivated by either of these considerations. Certainly the value of the Olympics as a means of boosting international exposure and inbound tourism was recognised, but this was perceived more in a national context rather than relating specifically to Sydney. In the previous two rounds of bidding for the right to host the Games, Melbourne had been Australia's officially endorsed, but unsuccessful candidate. Looking ahead to the 2000 Games, Sydney was quite simply, and pragmatically, seen as having a much better prospect of winning the bid for Australia. This success inevitably led to a number of significant planning and development responses within Sydney as both state and local government sought to deal with the challenges of both staging the Games and its longer term legacies.

The Olympics certainly provided an opportunity to resolve a longstanding urban development problem. Homebush Bay, the nominated site for the major venues, was a blighted, contaminated and largely redundant industrial wasteland located at the geographic centre of the City. Amongst its former uses were an abattoir, a brick pit, a munitions factory and storage site, and a number of chemical dumpsites. The Games provided the impetus for the area's rehabilitation and redevelopment. Prior to the successful bid, the state government had prepared a masterplan for the area and created the Homebush Bay Development Corporation (HBDC) in 1991 to oversee its implementation. The need to accommodate the Games was incorporated into the masterplan and HBDC's responsibilities were subsumed by the Olympic Coordination Authority (OCA) in 1995 (Sydney Olympic Park Authority 2008).

The early 1990s was also a time when major new markets were emerging from the so-called 'tiger economies' of Asia, with inbound arrivals from Korea and Taiwan growing particularly strongly (Griffin & Darcy 1997). By the middle of the decade the oversupply in Sydney hotel rooms had been absorbed and a concern was arising over whether the City would be able to accommodate the expected influx of visitors during and after the Olympics. The Tourism Olympic Forum, a body created by the state government to advise on tourism related matters in the lead-up to the Games, highlighted this prospective shortage in a report it commissioned soon after its establishment (Hotel & Tourism Asset Advisory Services and JLW Transact 1994). The Sydney City Council moved swiftly to stimulate more hotel development by amending its local planning controls. Floor space ratio bonuses, allowing up to an additional 40 percent above the normal maximum permissible floor space, were introduced for hotel developments in the CBD, provided the development application was submitted by 1 January 1998. This was specifically designed to stimulate development that would be ready in time for the Olympics.

The City Council and the NSW government recognised that there was a great need for public as well as private development activity if Sydney was to successfully host the Olympics. Aside from the facilities directly associated with operating the Games, Sydney's public domain, especially in its central areas, was perceived to require a substantial facelift. A major programme of civic improvements was initiated, particularly focusing on the CBD and key harbouside precincts such as Circular Quay. As a culmination and an ongoing legacy of this process, the state government created the Sydney Harbour Foreshore Authority (SHFA) in 1998. Immediately, SHFA assumed control of The Rocks from the Sydney Cove Authority.[3] The Darling Harbour Authority was also to be abolished and its responsibilities absorbed by SHFA following the Olympics. Hence, by 2001, one single government authority was landlord, planner, manager and promoter for Sydney's two most significant tourism precincts. SHFA was also given management powers over other important waterfront precincts such as Circular Quay and Woolloomooloo. As with all of its antecedent bodies, the main motivation behind creating such a powerful authority as SHFA was to ensure the dominance of state over local government in relation to the planning and development of these areas.

The decade had not been one of continual smooth growth, however. Some markets had waxed and waned. The Asian financial crisis in 1997 sent previously booming inbound markets like Korea into sudden decline. One market that grew steadily and significantly over the decade, however, was the backpacker market. Moreover, backpackers were having a profound effect on the geography of tourism in Sydney. Rather than the CBD and immediate surrounds, backpackers were being accommodated in areas that offered night-life and entertainment, such as Kings Cross and other inner suburbs like Glebe and Newtown. The beachside suburbs of Bondi, Manly and Coogee were also particularly targeted. From a developer's perspective, these areas had the added advantage of an ample supply of low-rent housing, especially boarding houses, that could be easily converted into cheap accommodation for backpackers. A problem with this trend, however, was that the backpacker lifestyle was not always compatible with that of the residential neighbour-hoods they now occupied. Local councils in the affected areas formulated development control plans to deal with many of these conflicts.

Backpackers aside, the bulk of Sydney's tourism remained concentrated around the CBD and adjacent harbourside areas. In the run-up to the Olympic Games, Darling Harbour, that was to be one of the main venues, strength-ened its position due to the completion of Cockle Bay Wharf on its eastern side, adjacent to the CBD. This area had been a void within the precinct for the previous decade. From a policy perspective, though, there was a grow-ing belief that Sydney needed to diversify its attraction base if it was going to maintain its competitiveness into the future. Hence Tourism New South Wales (TNSW), the state government authority responsible for tourism, launched first an attractions development strategy for Sydney (TNSW 1996) followed by the Sydney Tourism Experience Development (STED) Program (TNSW 1998). The STED Program in particular was focused on assisting the development and promotion of tourism in parts of Sydney outside the centre. One of the long term legacies of the STED Program was that TNSW established a number of committees comprising precinct managers and/or business interests, that were intended to maintain the impetus towards diversifying tourism experiences in Sydney. However, the potential of these committees to have a substantial impact on the quality or type of experi-ences available to tourists has never been fully realised.

Beyond 2000: Recovering Lost Ground

The new millennium could not have begun on a higher note for Sydney. The Olympics were acknowledged as a resounding success – the 'best ever' according to the IOC President – and now the City's tourism industry could sit back and reap the rewards from the enhanced international profile. The Sydney Olympic Park Authority (SOPA) had been established in July 2001 to take over the ongoing management of the Homebush Bay Olympic site and to assist the state planning authority with the area's conversion from a one-off venue into an integrated part of the urban fabric (SOPA 2008). This was not without its challenges, and the government recognised that to max-imise the tourist potential of the site it had to generate a sense of vitality around the somewhat monumental venues and not just rely on the staging of events. The area needed to be a living place with a substantial, permanent residential community.

All the post-Olympic optimism, however, came to a shuddering halt early in the morning of 11 September 2001, New York time. The impact of the terrorist attack on the World Trade Center on international tourism was felt worldwide, and Sydney experienced a decline in visitation like many other destinations. Two years later, the SARS virus emerged to further impact on a number of major Asian inbound markets. Talk of the positive Olympic legacy was now replaced by discussions of crisis management and recovery programmes. The expected post-Olympic boom never happened and the Sydney tourism industry became more concerned with consolidating and holding onto the gains that had been made in the previous decade. At least backpackers were still arriving in increasing numbers, and there was rapid growth from new Asian markets such as China and India. There was uncertainty, however, about how to best cater for these markets and, more importantly, how to avoid the problem of 'profitless volume' that was evident in the burgeoning Korean and Taiwanese markets of the 1990s. Of most concern was the decline in the Japanese market, one of the mainstays of Sydney tourism for the previous two decades. Furthermore, this trend could not be attributed to any crisis. Could it be that Sydney's appeal to tourists was on the wane, and it was no longer generating positive word-of-mouth and strong repeat visitation from its traditionally strong markets?

At the same time, tourism in Australia's second city, Melbourne, was thriving on the back of some extremely successful promotional campaigns. Melbourne was being portrayed as a sophisticated, cultured city that offered a diverse array of interesting and intimate experiences rather than iconic sights and attractions. Melbourne offered the prospect of discovery, whereas Sydney had perhaps become a little too well known and familiar. In the face of this success by its southern neighbour, in the latter part of the decade Sydney was seen by many as losing its competitive edge. Arguably it needed to learn some lessons from Melbourne's success and to focus more strongly on extending the range and quality of experiences the City offered. Sydney could still offer peak experiences, such as the Harbour Bridge climb, and its major and famous precincts like Darling Harbour were thriving, but would this be enough to ensure its sustained success?

An examination of recent trends in Sydney's tourism produces some conflicting messages. Domestically the number of overnight visitors to Sydney has decreased from 8.3 million in the year ended June 2003 to 7.6 million in the year ended June 2007, a decline of more than 9 percent over that five-year period. Domestic visitor nights have declined by more than 5 percent over that same period, while day trips have declined by 7.5 percent. Running counter to these trends, international visits and visitor nights grew by 16.5 and 37 percent respectively from 2003 to 2007. While the international visitation figures seem positive, it should be stressed that using 2003 as a base year may produce a somewhat misleading picture, given that events such as the SARS outbreak contributed to a significant decline in international tourism to Australia in that particular year[4] (TNSW 2007b).

Despite some concerns about the current trajectory of tourism in Sydney, the City maintains its position as Australia's pre-eminent destination, and tourism has become a significant part of the local economy. For the year ended June 2007, domestic overnight visitors spent in excess of A$3.5 billion within the City (i.e. excluding cost of transport to Sydney) (Tourism Research Australia 2007b). By virtue of their longer average length of stay, international visitors spent more than A$5 billion in the same year, that represented one third of the total international tourism expenditure in Australia (Tourism Research Australia 2007a). Domestic day visitors contributed a further A$1.6 billion to the City's economy (Tourism Research Australia 2007b). It may or may not be significant that on average both domestic and international visitors now spend more per night during a stay in Melbourne than they do in Sydney. The average length of stay for international visitors in Melbourne also exceeds that for Sydney (Tourism Research Australia 2007a).

THE SYDNEY TOURISM EXPERIENCE

One of the explanations for the decline in domestic visitors and slower than anticipated growth in international tourism, has been a concern that Sydney has linked the development of tourism generally, and the City's image in particular, too closely to its harbourside attractions. According to TTF Australia (2007a), the City needs to innovate and create new experiences and attractions to encourage visitors. They note that while marketing campaigns create a certain 'buzz' about a city visit, they have to be supported by a distinctive and compelling tourism offering (TTF Australia 2007b). By implication, the Sydney tourism experience may be failing to match its marketing message.

The experience of the City has certainly been subject to recent criticism (Munro 2007 and Nixon 2008). While Sydney presents a visual panoply to visitors, Munro argues that:

> Sydney is a trophy wife. Like her smug husband, we bask in the glory of association and smooth over the rough spots. Sydneysiders struggle with their glamorous, sparkling city. Catch the bus across town, but not if you are in a hurry. A brisk walk would probably be faster.
> Walk a few bus or train stops to lose weight. But rug up against the wind tunnels and the deep shade thrown by our skyscrapers. Meet a friend for a drink in town. But do not expect to hear each other talking.
> (Munro 2007, p. 1)

A walk northward to the harbour along Sydney's main thoroughfare, George Street, lays bare the clatter and bustle of the City noted by Munro (2007).

Concrete and glass towers dominate the skyline. At ground level, pedestrians compete with buses and cars on narrow and congested streets. What does this cacophony of architecture, people and commercial enterprise tell us about Sydney? The answer is brief – very little really. It seems that many post-modern cities of the 'New World' are largely distinguished by their lack of distinctiveness. Indeed as Bridge and Watson (2000) observed, these cities typically have more in common with each other than with their surrounding hinterland!

As George Street approaches the harbour and the tourist hub of Circular Quay, visitors share the public space with city workers who scurry from the commuter ferries toward their offices. At this point a large steel railway bridge dissects the public space and fractures the flow of the skyline. Painted along the horizontal beam of the bridge is a sign 'Welcome to The Rocks'. Passing under the bridge, the physical and social characteristics of the cityscape are transformed. Twentieth century concrete and glass give way to nineteenth century sandstone and slate. The goal-directed pace of the City gives way to more relaxed, seemingly aimless activity. Predominantly commercial activity is geared to meeting leisure and pleasure needs. The collective effect of the change in architecture, scale, tempo and visitor activity, creates a markedly different city experience – this is clearly a different site. It is part of the City, yet a place set apart.

The Tourist Precinct

Precincts such as The Rocks play an important role in the life and the experience of the City for both visitors and residents. As noted elsewhere (Hayllar *et al.* 2008), tourism services and attractions, from both supply and demand perspectives, are not dispersed evenly and seamlessly throughout a city but tend to be concentrated into relatively small, contained geographic locales – precincts. For visitors in particular, precincts are focal points and sites of intense consumption.

In an 'intensely urban' city like Sydney, the interaction of tourists with precincts takes on a particular resonance. Arguably, precincts are counter-points to the dissonance of the City – they bring legibility to the discord that surrounds them. This position raises two interrelated questions about precincts that have in part been broached by both Maitland and Newman (2004) and Montgomery (2004). The first concerns how well do we understand the experience of tourists in precincts? What is their experience, how do they explain it, what sense of meaning do they attach to the experience? The second, and interconnected question, concerns the functional aspects of precincts. What functions do precincts perform in shaping or facilitating the experience of tourists? Can seemingly dissimilar precincts perform similar functions? Do inner-city precincts perform different functions to city fringe, or suburban precincts? The focus is thus on the roles a precinct performs and not simply what its physical form or 'type' might imply.

The chapter now turns to addressing these and related questions in respect of Sydney. The first part examines the experience of tourists in the two city-based precincts previously discussed, The Rocks and Darling Harbour, and theorises the role these precincts play in providing experiences beyond the well trodden path. The second section broadens the scope of the discussion and considers the functions these two precincts perform for tourists and contrasts them with two suburban or 'fringe' precincts – Norton Street (Leichhardt) and Parramatta. The overall purpose in this latter discussion is to examine the functional roles of precincts and their implications for the tourist experience of Sydney. Finally, and against this background, the future of Sydney as a tourism destination is considered.

PRECINCTS AS PART OF THE SYDNEY EXPERIENCE

The experiences of people in precincts are at least partly shaped by the historo-physical context of the precinct itself. A visitor's prior knowledge, impressions passed to them by others, and the physical fabric and location of a precinct are among a number of factors that impact on the expectations of visitors and, as a logical corollary, the experience itself.

The Rocks precinct takes its name from the rocky outcrops, long since flattened by development, that once marked its shoreline and immediate environs. In earlier times, The Rocks had a reputation for its hard working-class edge and nefarious activities such as those perpetrated by the *Rocks Push*, a street gang which terrorised the area in the late nineteenth century.

In perhaps a portent of its future as a tourist precinct, an Englishmen visiting Sydney in *circa* 1890 was taken by a friend into a Rocks pub frequented by *Push* members and described them as:

> Wiry, hard-faced little fellows, for the most part, with scarcely a size-able man amongst them. They were all clothed in 'push' evening dress – black bell-bottomed pants, no waistcoat, very short black paget coat, white shirt with no collar, and a gaudy neckerchief round the bare throat. Their boots were marvels, very high in the heel and picked out with all sorts of colours down the sides.
>
> (A.B. Paterson – *Push Society* 1906)

Although the *Push* has faded into history, many tourists still venture into The Rocks in an attempt to capture a sense of the 'real' Sydney. While substantially altered and 'sanitised', The Rocks offers tourists the prospect of insights into the City's past and present day contemporaneously and has become a significant focus of city-based tourist activity. Between July 2006 and June 2007 the precinct recorded over 13.5 million visits. Sydney residents are the dominant user group (54 percent) followed by international tourists (24 percent) and those from regional areas of NSW and interstate visitors (22 percent) (www.shfa.nsw.gov.au).

Approximately three kilometres along the harbour foreshore from The Rocks is the contrasting precinct of Darling Harbour. Conceived and developed in the genre of the 'festival marketplace' (Rowe and Stevenson 1994), Darling Harbour is an eclectic mix of modern architectural forms reflecting its twenty-year development 'heritage'. Despite some ongoing criticism (see Huxley 1991, Hall 1998 and Searle 2008) it has become a weekend hub of activity for Sydney residents and visitors alike. The regular events, array of attractions and opportunities to stroll around the waterfront, all with the backdrop of the City and harbour, make Darling Harbour an attractive leisure time option. Between July 2006 and June 2007, the precinct received over 27.5 million visits making it Sydney's most visited tourist place. Of these, 62 percent were Sydneysiders, with the remaining 38 percent divided evenly between domestic and international visitors (www.shfa.nsw.gov.au).

As Sydney's most significant tourism precincts, The Rocks (Hayllar and Griffin 2005) and Darling Harbour (Hayllar and Griffin 2007) provided the foundation for the first two in a series of studies undertaken by the authors. The overall purpose of these studies was to progressively develop an under-standing of tourists' experience of a city generally, and in particular their experience of city precincts. In each of these studies a hermeneutic phenom-enological approach to the examination of experience was used. Data were collected via in-depth interviews using the respective visitors centre in both precincts to recruit participants and conduct the interviews. Interviews were conducted with ninety visitors; sixty international and thirty non-Sydney residents. Verbatim transcripts of the interviews were prepared and the narrative 'data' were analysed using the thematic approach recommended by Van Manen (1990) and Denzin (1989). Following several 'works' of the participants' narratives by the research team, three experiential themes emerged from the data:

- physical form and setting;
- atmosphere; and
- personal meaning.

The *physical setting*, and what it means to the visitor, is an important experiential characteristic. The Rocks and Darling Harbour have strikingly different architectural forms. The historic yet evolved streetscape of The Rocks with its low rise, more human scale eighteenth and nineteenth cen-tury buildings links Sydney to its colonial past. In contrast the 'style' of Darling Harbour is somewhat of a discordant mix of late twentieth century concrete, glass and metal. In both cases the precincts are located close to the City yet they eschew quite a different experience. The Rocks is adjacent to the main tourist core of the City. It is encircled by vibrant inner-city life and tourist activity. Its relative quietness and change of pace are in sharp contrast to its immediate surrounds. It is a place to meet, eat, drink and shop – it is a more urbanised, perhaps local, experience. By way of contrast, *Play it your way* is the marketing theme for Darling Harbour. It is a site of colour, movement and activity. As one respondent in the Darling Harbour study commented: *It's like a big entertainment centre.*

The second theme, *atmosphere*, refers to the overall 'feel' of the precinct created by the dialectical interaction of the social and personal experiences of the visitor with the precinct. A visit to a precinct is an inherently social experience. Both Darling Harbour and The Rocks are urban melting pots of international and domestic tourists, local residents and office workers. On weekends, the office workers make way for pleasure-seeking locals and visitors from outside Sydney. The atmosphere in The Rocks is more sub-dued, 'old world' in style, while Darling Harbour is more open, expressive and fast-paced. The Rocks' narrow streets, living precinct character and human interactions convey a particular type of experience; in part a sense of timelessness. Darling Harbour 'feels' younger, more active and a place of its time. These different characteristics help shape the personal experiences of the visitors.

The final theme identified is concerned with how an individual's cumula-tive experiences and level of engagement provide some sense of *personal meaning* to their visit. In the sense described here, meaning is understood at two levels – external and internal. External is concerned with the meaning the precinct gives to the 'host' city itself – as a marker for Sydney as a destination. Conversely, internal meaning is more inwardly focused – what the experience means for the individual.

In both precincts there is an unambiguous sense that visitors are experi-encing 'Sydney' but in manifestly different ways. Both are measures of dis-tinctiveness. At The Rocks, visitors recognise they are moving in the environs of old Sydney and do not appear to be concerned with what might be con-sidered the imposition of the modern or 'inauthentic'. Indeed Wang's (1999) notion of 'existential authenticity' (in Ashworth and Tunbridge 2000, p. 17), where authenticity lies in the experience and not the object, is relevant to this group of visitors. Conversely, Darling Harbour provides a different marker. Here the experience is one of modern, emblematic Sydney.

Respondents at both sites expressed positive 'feelings' toward their experi-ence. They felt 'safe', 'comfortable', 'less stressed', and 'not hassled'. However, these expressions were more marked in reference to The Rocks. It could be that the greater 'depth' and complexity of the environment in The Rocks provided a richer experiential foundation for meaningful reflection. Indeed support for this observation is provided in companion studies of precincts undertaken in Melbourne (Griffin *et al.* 2006).

Theorising the Precinct Experience

In thinking through the narratives and thematic analyses for both these precincts as described above, we have argued from the phenomenological perspective that the contested notion of *place* (see Relph 1976, Lefebvre 1991, Suvantola 2002 and Creswell 2004) is the essential characteristic or *essence* of the visitor experience to these spaces (Hayllar and Griffin 2005). While the companion concepts of *space* and *place* are problematised in the literature, the work of Suvantola (2002) and Tuan (1977) is instructive. According to Suvantola (2002), there are five different types of spaces: mathematical, physical, socio-economic, behavioural and experiential. It is his notion of experiential space that is of particular interest. Suvantola (2002) argues that experiential space is dynamic and, invoking phenomenological reasoning, describes the use of space as both lived *and* experienced. Of the five different types of spaces discussed, it is perhaps the least quantifiable yet arguably the most important in respect of the tourist experience. It is within the experiential realm that meaning is applied to space through our encounters with it. Theoretically, a space imbued with meaning becomes a place (Tuan 1977).

Although Darling Harbour and The Rocks are sites of substantial contrast, they each have an intrinsic resonance that visitors connect to Sydney. The unique architectural heritage of The Rocks, and its location as a physical and social counterpoint to the City which adjoins it, makes this a distinctive urban landscape in Sydney. It is this distinctiveness, and the visitors' experience of it, that give rise to phenomenological experience of place.

The phenomenon of place emerged somewhat unexpectedly from the analysis of Darling Harbour given the pejorative statements that generally accompany discussions of such planned precincts. For example, Rowe and Stevenson, argued that 'festival marketplaces involve the calculated packaging of time and space, seeking to satisfy tourists' expectations of an authentic experience of place by constructing often decontextualized and sanitized simulations of urban landscapes', and further that these types of precincts resonate with the 'urbanism of universal equivalence so that anywhere can now be everywhere' (Rowe and Stevenson 1994, p. 181 and see also Clarke 1998, in Craig-Smith and Fagence 1995). However, our findings suggest that the visitor experiences of Darling Harbour run counter to this conventional wisdom. Darling Harbour is not 'anywhere' nor simply the 'carnivalesque' with its implied superficiality and depthlessness, but rather something more engaging and meaningful for the visitor (Hayllar and Griffin 2007). It is a distinctly Sydney place.

In considering the overall thematic analysis, and the implicit psychological shift from space to place, it could be argued that there is a type of experiential or thematic 'hierarchy' in the nature of the visitor's interaction with the precinct. This hierarchy is ordered by increasing levels of engagement or depth in the way a precinct is experienced. At the broad level, the *Physical Form* is immediate and unavoidable. At one level it may invoke the spirit of curiosity and invite further exploration – to investigate what is beyond the immediate or to discover 'what's this all about?' At another level, the physical form may act as a barrier to further engagement. Where there is no experiential (emotional) union between the physical form and the visitor, arguably the precinct remains a space rather than a place.

The next level, *Atmosphere*, relates to the social psychological experience of engagement with the precinct beyond the more immediate reaction to its physicality. It is an emotional sense of experience which we argue is critical for the precinct to be experienced as a *place*. Thus emotional commitment in its psychological sense leads to an experience of greater depth and subsequently meaning.

Finally, and consistent with Tuan's argument, the third level, *Meaning*, represents the deepest level of experience, beyond that of immediate enjoyment. At this less superficial level of experience, the visitor acquires a much stronger sense of place attachment. Indeed it might be that a 'meaningful' precinct experience may have a substantial impact on the overall destination's sense of place much like landmark sights have in other destinations, for example the Eiffel Tower in Paris or the Statue of Liberty in New York.

A second set of theoretical considerations concerns the experiences provided by the study precincts for different types of visitors. In a previous paper (Hayllar and Griffin 2005) we discussed a typology of visitors identified as *Explorers*, *Browsers* and *Samplers*. This typology reflects the idea that different people can simultaneously experience a precinct in quite different ways, provided the precinct offers opportunities for different layers of experience.

The *Explorers* were identified as those visitors who want to move beyond the façade of a precinct, to find their own way and discover its innermost complexities and qualities. An explorer in The Rocks who wanted to look beyond the immediacy of the precinct commented that:

> *The narrow cobbled street is something you don't find in very many places and I think that's really part of its charm. You're walking along one street and then all of a sudden without realising it there's a kind of this little alleyway off to the left that might take you to something a bit more interesting around the corner.*

Another noted that *you can just kind of explore behind the buildings and through the alleyways. It's fun. It's got wonderful energy.* Earlier she commented that if you wander *you never know, well you know you've got to come out somewhere, but I like the little streets.*

The second group, the *Browsers*, are more content to stay within the confines of the main precinct area and to follow the tourist routes. As one browser commented: *you just want to sit, listen to the music* or call into the local pub *to mix with people a little bit and that's quite nice.* Another browser considered The Rocks as a collection of

> *kind of quaint little restaurants, little delis, little cafés – very relaxed, very casual and something you'd want to go and see if you just came off the boat or something [if] you were a little bit tired but didn't want to go home yet. It's a nice place to stroll around to relax and take a break.*

The browsers are engaged by the experience at hand but do not seek to further enrich that experience through extensive investigations of the precinct.

The experience for the final group, *the Samplers*, is more superficial, with a visit to a precinct representing just another stop on their schedule of moving through the attractions of a city. For them the precinct is yet another sight to be viewed but not experienced, or a place of brief respite. The sampler will typically not move beyond the fringe or specific 'refuge point', such as a café.

Taken together, this theorising around the precinct experience raises three key points: firstly, the psychological shift from the experience of space to place as described by Tuan (1977) is arguably a consequence of the dialectical interaction between the physical and social experience of the precinct itself; secondly, that the depth of engagement with a precinct may be hierarchically arranged; and finally, that the type of experiences sought from and the level of interaction with the precinct may be dependent upon the learned predispositions of visitors for experience seeking behaviour.

In the context of the broader discussion, arguably it is the explorers who actively seek experiences away from the mainstream tourist environs of a city. It is their desire for exploration that motivates them to take the path less trodden. When a city precinct has been fully explored, it is the explorer, who may also be a second or third time visitor to the City, who seeks more meaningful and place engaged experiences.

SYDNEY PRECINCTS AND THEIR FUNCTIONS

Taken together, the thematic constructs and theorising point to the multi-faceted nature of a city experience where physical structures, activity, human interactions and personal experience engage in a vital interplay that produces an 'experience' for the visitor. Interestingly the visitor interviews in The Rocks and Darling Harbour also highlight how two physically disparate and contrasting precincts can fulfil similar roles or functions for the visitor. For example, both: provide points of contrast or act in the role of a 'refuge' or place of respite from the City; provide opportunities for interaction with locals – they are part of the social fabric; and contribute powerfully to a sense of identity for Sydney.

The development of this 'functional' reasoning led to a second set of studies, the principal purpose of which was to better understand the functions that precincts perform for the visitor to an urban destination. Hence the focal question of the study moved from 'What is my experience in this precinct?' to 'What functions do precincts perform, for me as a tourist?'. A related supplementary question was to examine, in part, the relative importance of these various functions. These follow-up studies were conducted in both Darling Harbour and The Rocks. However, in an attempt to move beyond the city-based sites, studies were also conducted at two precincts outside of the immediate central-city boundaries: Norton Street in Leichhardt, and Parramatta.

Norton Street would be typically be classified as an 'ethnic' precinct. Leichhardt is an inner western suburb situated eight kilometres from the Sydney CBD and is readily accessible to the City via light rail and bus. In the 1920s Italian migrants began to settle in the area, and this increased dramatically in the years following World War II. Hence Leichhardt became substantially influenced by Italian culture, particularly with the appearance of Italian businesses and community clubs along Norton Street. The Italian/Australian newspaper, *La Fiamma*, was also established during this period. Today, the *Italian Forum* shopping and residential complex, the *Casa D'Italia* – headquarters for an Italian cultural and social support agency – cafés, restaurants, cinema and shopping experiences in suburban Norton Street retain a distinctly Italian feeling and flavour (see Figure 6.4).

The second precinct, *Parramatta*, was founded in 1788, the same year as the original British settlement of Sydney. Located twenty-three kilometres from the CBD, Parramatta is now at the geographic centre of metropolitan Sydney. Although accessible by both train and bus, the most popular entry point for tourists is by the *RiverCat*, a fifty-minute ferry ride from Circular Quay along the historic Parramatta River. Famous for its sandstone Georgian architecture, Parramatta features some of the most important historical sites and buildings in Australia, including Old Government House (Australia's

Figure 6.4 Sydney: The Italian Forum Norton Street
Source: Authors' own.

oldest surviving public building), Experiment Farm (the site of Australia's first wheat crop), Elizabeth Farm (the home of John and Elizabeth Macarthur – founders of the Australian wool industry), and the Female Orphan School. While Parramatta presents itself as a unified precinct, in reality these important buildings and sites are widely dispersed throughout the area in 'micro' precincts and are typically difficult for all but the most intrepid tourist to access. Otherwise Parramatta represents a major commercial and retail centre, regarded as Sydney's second CBD serving its extensive western suburbs.

Data were collected through a questionnaire survey administered within the four precincts indicated above. The questionnaire design was based primarily on ideas which unfolded from interviews with precinct managers and previous qualitative work (Hayllar and Griffin 2005 and 2007). The same questionnaires were administered in each of the four study precincts, allowing for the direct comparison of the results.

In the course of the study, 690 domestic and international tourists were interviewed at the four sites. Essentially the questionnaire was based on presenting the tourists with a wide range of statements that each reflected a particular function that the precinct could be performing for them. The respondents were asked to indicate their level of agreement with each statement, based on a five-point Likert scale. For the purposes of analysis presented here, only eleven (of thirty) of the most highly ranked functions are reported. While the overall analysis is still at its preliminary stage, the data show that the highest ranked functions (based on mean scores) group themselves around three functional themes: social functions, atmospheric functions, and facilitating functions.

In the context of this study, *social functions* are concerned with the extent of the visitors' engagement with the precinct. The sense of personal freedom and comfort to move around with minimal social constraint appears important to visitors. This was particularly the case in those precincts where tourists typically gather in larger numbers and where tourist related services appear as part of the natural landscape – The Rocks and Darling Harbour. Visitors to three of the four precincts (Parramatta being the exception) considered the opportunity to sit back and observe the experiences of others – 'people watching' – as important to their experience. However, there appear to be some mixed messages, perhaps sensitivity expressed through the data, as visitors did not want to intrude in the lives of Sydney residents. This was noticeable in the suburban precincts of Parramatta and Leichhardt where their presence as 'outsiders' is more readily identified. This finding presents a paradoxical dilemma for the explorer who wishes to investigate sights off the beaten track but does not wish to stand apart from the everyday experience of the locals.

In the social context, precincts function to provide a 'space' for social engagement somewhat at arm's length from the resident. Seemingly visitors want to experience the sociality and freedom that these spaces provide – to sit back and observe others – while enjoying the autonomy of the visitors' space.

The *atmospheric functions* that emerge from these data reinforce the importance of a precinct's atmosphere as part of the overall visitor experience. All precincts provided a 'relaxing atmosphere' for their visitors. Some earlier theorising around the notion of precincts performing the function of a refuge (Hayllar and Griffin 2005) from the City is well supported by these data. This was particularly the case in The Rocks, and Darling Harbour but less so in Norton Street and Parramatta. The result is perhaps not too surprising given the more suburban nature of the latter two precincts in contrast to the holiday atmosphere engendered by the more iconic precincts.

Visitors' responses to a related question on the 'vibrancy' of the atmosphere in particular precincts appear to be consistent with our observations and indeed the marketing of the precincts. Darling Harbour ranked the highest on this atmospheric function. The more laid back atmosphere eschewed by both the historic precincts – The Rocks and Parramatta – and the suburban Norton Street are borne out by the data.

Visit *facilitating functions* refer to the specific interventions or planning decisions taken by managers, such as wayfinding or lighting for evening use, that in some way facilitate or shape the experience of visitors. As expected, the more developed precincts – Darling Harbour and The Rocks – were ranked highest in this functional area. A related visit facilitating function that scored consistently across each of the four precincts was their importance as a point of reference to navigate to and from other parts of the City.

While not specifically commented upon in the data, meaningful information on the suburban precincts and their links to adequate public transport is problematic. At the Parramatta ferry wharf, there are no tourist services to facilitate movement between the micro-precincts, nor is adequate wayfinding provided. Thus even though tourists visit Parramatta with intent, the infrastructure needed to facilitate the experience, even for the most dedicated and adventurous, is conspicuously absent. The data suggest that Parramatta acts more as the end of a ferry excursion up the Parramatta River rather than living up to its potential as an opportunity for exploration into Sydney's history.

Norton Street is less impacted by transport problems. However this small living precinct may be constrained by the depth of experience it can offer to visitors. It takes time to access the site for what might seem quite a limited experiential return. To this end the Norton Street precinct managers likely need to consider how they can project its image through interpretation and maximise its potential.

Experience and Functions – The Sydney Visitor Experience

The majority of tourists to Sydney visit its two principal precincts – The Rocks and Darling Harbour. Here, they have the opportunity to seek 'refuge' from the City, yet at the same time engage with it through its architectural forms and social interactions with locals. In these geographically small areas, tourists form impressions of Sydney as a place. Indeed for many short term visitors, this is Sydney. They give Sydney meaning – one historical, one contemporary. They are places where Sydneysiders predominate and where visitors can observe and engage with anonymity because everyone within these areas is a visitor (save for a small resident population in The Rocks). The transient nature of visitors to these precincts creates an atmosphere which 'feels' comfortable. Visitors can be with others in the overtly social context of these precincts, while at the same time having personal space and opportunities to gather their thoughts through reflection on their own experiences – time *away* from, and time *with* others.

In experiencing these precincts, it is clear that they perform a number of functions for tourists – many of which have been facilitated by design and management considerations or have simply evolved from the existing fabric. The important social functions, identified across all precincts, suggest the need for spaces for both open engagement with others, and spaces for respite. This may be catered for in the public domain, such as the open waterside promenades in Darling Harbour or through the specific commercial uses of space, such as the back street cafés and pubs in The Rocks.

The design and use of these spaces, the people flows, and the interactions with the physical structures, help create a distinctive atmosphere in each of the precincts. The facilitating functions, such as places of orientation (obvious landmarks or cultural meeting points) or designated wayfinding mechanisms like maps or directional signage, serve to move or guide visitors through their experience.

On the fringe of the City, Norton Street and Parramatta are geographic 'outliers'. While clearly they perform many of the functions identified across the precincts, they do not perform as strongly relative to the other precincts – Parramatta in particular. Parramatta performs above the mean on only two of the eleven most highly ranked functions identified while Norton Street is more successful, exceeding the mean scores on six of the functions.

Overall, the well known precincts are performing their expected functional roles and generally providing the types of experience valued by tourists. However, it would appear that the suburban venues are not attracting the number of visitors, meeting their functional needs nor realising their potential to provide opportunities for a deeper engagement with Sydney by particular groups of experience seekers.

In the context of earlier discussions, arguably it is the explorers who might venture to the suburban precincts given their need to seek experience beyond that which is presented to them within the typical tourist frame. However, if the suburban precincts are not meeting the functional requirements and, *ipso facto*, the explorer's need for a particular type of experience, these precincts are not contributing in a meaningful way to the depth of tourist experiences offered by the City – they are off the beaten track and will continue to be so!

SYDNEY TOURISM: THE FUTURE

In this chapter we have argued that the tourists' experience of Sydney is substantially influenced by their interaction with its iconic harbourside attractions. However, in moving beyond the immediacy of this initial experience, the two most visited precincts in Sydney, Darling Harbour and The Rocks, provide visitors with the opportunity to better understand the City, albeit in somewhat contrived environments created or managed largely for tourists. Here visitors interact with locals and other tourists, and here they form early and sometimes lasting impressions of Sydney through their physical and social interaction with people and space.

While for many tourists these sites provide the opportunity to explore 'Sydney', for others, their explorations are limited by both time and accessibility. The CBD itself is congested and unexceptional. A ferry trip to Manly or a bus ride to Bondi complements the City experience of many visitors. (Indeed, both these beachside suburbs have been substantially influenced by backpackers, who have been attracted by the seaside locations and quintessentially Australian experience that these locations are seen to offer.) However, Sydney has the potential to offer far richer and more complex experiences for the visitor, in keeping with its status as a 'world city'. In extrapolating from the findings relating to the fringe precincts, it could be argued that Sydney is currently falling short of realising that potential; the number of visitors to these precincts is relatively small, and the experiences they offer to those who venture beyond the central-city are not remarkably rewarding.

The question arises as to whether any of this matters. Regardless of the extent to which tourist currently explore the City, Sydney is still a successful tourist destination in aggregate terms. It is still the most visited destination in Australia, receiving some 7.4 million domestic visitors (Tourism Research Australia 2008b) and 2.7 million international visitors (Tourism Research Australia 2008a) in 2007. It retains its pre-eminence over Melbourne, with 13 percent more domestic visitors (Tourism Research Australia 2008b) and nearly twice as many international visitors (Tourism Research Australia 2008b). Of some concern for tourism interests in Sydney, however, the gap is narrowing, particularly in relation to domestic tourism. Indeed in 2007 the total expenditure by domestic visitors to Melbourne (A\$3.39 billion) was only marginally behind that of Sydney (A\$3.45 billion) by virtue of Melbourne visitors' higher average expenditure per night (A\$179 compared to Sydney's A\$158) (Tourism Research Australia 2008b).

The policy development and marketing of Melbourne as a destination may well be instructive. With no 'world city' pretensions and limited landmark attractions, Melbourne projects itself as a type of metaphorical theatre – the events 'capital' of Australia – and a sophisticated city to be enjoyed and explored. Melbourne embraces the explorer. Its promotional campaigns celebrate difference and complexity and invite the visitor to experience this difference; typically through engagement with thematically organized precincts spread throughout the City. Moreover these campaigns suggest that Melbourne is a place that evokes emotions, particularly romance, and engenders a sense of attachment, with one series of recent advertisements suggesting 'You'll never want to leave'.

In Melbourne, the notions of experience and opportunities for exploration are central to its positioning. By way of contrast, the promotional tagline that 'There's no place in the world like Sydney' (TNSW 2007a) continues to draw on its established attractions as its key marketing message. The predominant message of Sydney's promotion seems to be 'We're Special' (it's about us), whereas Melbourne's suggests 'We'll Make *You* Feel Special' (it's about you). In our view, Sydney's sustained success and maintenance of market share in the face of growing competition may rely on conveying the idea that it offers distinctive and personally enriching experiences beyond the well known sights. Indeed TNSW policy documentation (2006) focusing on 'Sydney Surrounds', supports such a proposition. However, it is unclear whether or not this is having an impact on visitation patterns to the areas beyond the central-city and harbour.

Within the CBD and its immediate environs, Sydney needs to better facilitate and articulate its rich heritage. One way to deliver on this offer is to adopt a Melbourne type approach, that itself has been modelled on successful international examples, and develop an array of accessible and logically connected precincts throughout the City that, by virtue of their form, activities and atmosphere, possess the qualities that facilitate the experiences that tourists seek. These precincts need to provide depth to accommodate the different visitor types (different experience seekers) and to encourage repeat visitation. They need to enrich the Sydney experience both within and beyond the immediate city – beyond iconicity.

7 Conclusions

Robert Maitland and Peter Newman

Visitors to cities share their appreciation of characterful places and good places to eat with residents and city workers. In the places the authors of these chapters have examined it is becoming more and more difficult to distinguish between visitors and others. We asked the contributors to look at both how tourism was changing in their cities and at the new places off the beaten track, that visitors were getting to. Their conclusions suggest some broad similarities among this group of well known, well visited and multi-functional world tourism cities, discussed below. In this final chapter we would like to explore the strength of some of these similarities and the implications of the findings for tourism research and for urban studies more generally.

VISITORS AND THE CITY

In the introduction we pointed to some of the weaknesses of available statistical data on tourism. Our contributors have drawn together key figures from their cities but exploring the similarities and differences in the experiences of the cities is made more difficult by the absence of common and comparable statistical data on tourism. This is a twofold statistical problem. First, comparatively little data is collected, despite the substantial and growing importance of tourism in all the cities. We know all too little about who visits the cities, what they do and what they like. What data is available tends to reinforce existing expectations and conceptions of tourism development. Visitors to recognised tourist attractions are relatively easily counted, and there is usually a strong motive for operators – whether commercial or not – to monitor their likes and dislikes. Keeping track of the other areas of a city that people visit, and what they enjoy is harder to do and no single agency has a responsibility or clear motivation to do it in a consistent and regular way. One result is to reinforce a perception that what matters are big attractions rather than the subtler appeal of exploring a city. In some ways this is not surprising for, as Fainstein (2007) points out in the case of New York, the hotel and convention bureau is supported by big Manhattan based hotels and has little interest in other localities off the beaten track.

The second problem with tourism statistics is comparability. Different cities collect different data based on different definitions, and assembling comparable information on trends and changes is made more difficult. More data that is more comparable would allow the cities to learn more from one another's experience.

However, there is more fundamental concern about how we understand tourism in these cities. Long established official definitions distinguish tourists from others in terms of time and distance (Michael 2008). Tourists are categorised on the basis of having travelled away from home, and being in a place only temporarily, with an implicit assumption that this will distinguish them and their demands from a separate host population. In many cases this is still reasonable enough – in resorts, rural areas and many tourist historic cities for example. In world tourism cities too, it still accurately reflects some visitors and their demands. Many arrive wanting to consume iconic attractions like the Eiffel Tower, Big Ben, the Reichstag, the Statue of Liberty and Sydney Harbour and to visit well known tourism precincts. However, in these cities familiar divisions between 'visitors' and 'hosts' are increasingly problematic. Distance travelled and time spent in a city do not in themselves tell us much about what visitors want or do. Rather, we need to consider the demands, behaviours and practices of a wide range of city users with different mobilities and different attachments. As the accounts from our cities show, these frequently overlap and attempts to separate 'visitor' and 'host' can obscure attempts to understand processes of change. There remains analytical value in considering the demands and experiences of those visiting a city but it is more helpful to see them as occupying one part of a range of mobilities and preferences rather than as a clearly distinct group. Demanding a different quality of information about city users also has implications for how we study the places they use.

Tourists are thought to look at things differently, to see places with a tourist gaze (Urry 1990), and to be curious to look at 'the other' because it is different from themselves. In the cities we have examined things are more complicated. Visitors are frequently experienced city users and often familiar with and attached in some way to the cities they visit. They want to fit in, rather than stand to one side. In Berlin it was clear that for those going to Prenzlauer Berg and Kreuzberg much of the appeal was to do with blending in to the social fabric, and establishing a role that was more performative and interactive. Gazing on 'the other' – punks or scenesters – is a part of that, as it is for city residents, but does not dominate the experience. In London, the appeal of everyday life drew visitors off the beaten track, and it was the presence of locals that marked out areas as places where it is possible to experience the 'real' city. In Sydney it seemed that a set of iconic attractions and the legacy from an Olympic Games widely regarded as a model for a successful event, were no longer enough to assure its leading role in Australia. Melbourne seems to be closing in, particularly in terms of high spending visitors since it offers not major landmarks and familiar icons but varied experiences and the opportunity to explore and engage with a diverse city. This suggests that Sydney – and other cities – need to find ways of offering enriching experiences beyond well known attractions. We will return to consideration the implications of the findings for tourism and urban policy makers.

However, tourists still do look at things differently. Whilst they want to fit in, it was clear across the cities we examined that some visitors were exploring and looking for the 'real city', places more 'typical' than what tourism precincts have to offer. Yet the real city is an elusive concept, and the places we examined are far from typical of the cities of which they form a part. These are multi-use areas, able to play a variety of roles and fulfil a range of functions. Whilst visitors feel they have found the real and the typical, we can speculate that longer term residents see these areas as rather special – places that offer the ingredients for a good day or evening out. There is no necessary contradiction here, but further research is needed to explore perceptions of visitors and – crucially – other city users. We need to understand better the perceptions and experience of all those who use both established tourism precincts and newer, off the beaten track areas.

TOURISM AND URBAN DEVELOPMENT IN WORLD TOURISM CITIES

We can conclude from this set of city studies that some types of urban visitor are increasingly consuming new experiences in attractive neighbourhoods close to more established tourist circuits. They tend to be more experienced urban visitors. They are looking for opportunities for consumption, to enjoy bars and restaurants but also to enjoy the less tangible qualities of distinctive places. The resident middle-class tends to share the urban preferences and consumption demands of experienced, middle-class visitors who may have business in the city, be visiting friends and relations or 'just' visiting. Let us now turn our attention to the cities and to a group of cities that seems to share the capacity to, apparently seamlessly, open up new areas for visitors to explore. We suggested that a world tourism city's advantages arose from the combination of leading economic activities, strong arts and cultural sectors, traditional attractions and, importantly, diverse residents and workers. These multiple and interlinked assets, and a polycentric urban form, create the conditions in which new tourism areas can emerge. There are similarities between the localities examined in this book – to greater or lesser degrees they share proximity to traditional tourism zones, or to financial or other functionally important districts, they share the accessibility of the public transportation investments of world cities and they offer new mixes of the cultural diversity and spatial forms of large cities. Each city offers the capacity to create new desirable locations and seems to confirm the competitive advantage of the world city assets. We may or may not find it helpful to see this similarity as yet more evidence of the homogenising force of contemporary globalisation where distinctiveness has become an essential asset in economic competition.

To be able to produce and reproduce tourism assets in this way adds to the already considerable competitive advantage of these cities. So there is an important consequence of our five city cases for public policy. We have seen how the particular places our contributors have explored have changed in recent years but also that they have changed without conscious tourism planning. Public policy has had important, but largely unintended, effects in developing tourism. In Kreuzberg, the long term development of strong neighbourhood politics enhanced and promoted its bohemian and alternative character. In London, long sought improvements to the overloaded Underground system brought new or improved stations and improved access. In Sydney, the 2000 Olympic Games were certainly planned for and affected tourism – but the policy intention was to promote Australia through the event; Sydney was the chosen venue because it stood a better chance of winning. Local initiatives in the New York boroughs, and the ability of some local actors to see how tourism might integrate with local regeneration objectives, were not translated into citywide policy. Cities may desire new middle-class visitors but there is no conscious effort in the places examined to plan for tourism. Perhaps something of an exception is the case of eastern Paris. In the plans for the regeneration of the Bercy warehouses and more generally in the remodelling of eastern Paris we find a more active style of intervention by city government, although the effects of the projects and in particular the attraction of visitors from out of town were certainly not perceived by the planners. Comparing Paris to the other cities we might perceive more conscious planning. Given the traditional perception of a dirigiste style of planning in France this difference of approach is perhaps not unexpected and the authors looked at the Paris case specifically through the lens of the urban planning project. But the difference between Paris and the other cities is one of degree and within the City the informal development of tourism in Belleville cannot be compared with the formal development of new attractions such as the Parc de la Villette, for example, and we can see change in eastern Paris as driven by urban renewal rather than tourism policy and planning. Policy processes differ among this group of cities. What they share is a recent recognition of the integration of tourism with other dynamics of change but without any full understanding of how these forces interact or any specific understanding of how to capture and reproduce the sorts of neighbourhood transformation we have examined. What links Belleville to Bankside and Brooklyn is not tourism policy but actors in search of the 'real' city and the unplanned, authentic urban experience. Visitors share this rediscovery of a residential suburb with the new class of gentrifiers moving east in the Paris housing market. It is this connection between tourism and wider urban processes that links all of the chapters.

FUTURE DEVELOPMENTS

Our World Tourism Cities have proved able to generate new areas for visitors to explore and to do so in a way that emphasises conviviality between different city users. However, our understanding of the process remains limited, although it does seem clear that it is not the result of deliberate tourism planning and policy.

One important question revolves around how far off the beaten track areas can continue to meet the demands of exploring visitors. Their appeal depends upon perceptions that they are comparatively undiscovered, and places where the locals go. It is unclear how robust this appeal will prove as visitation continues. City marketers are always eager to promote new attractions, yet areas that are promoted as off the beaten track inevitably lose that status. Even without such intervention, perceptions may change as visitor numbers grow, and areas adapt to seize market opportunities. These areas may be unstable (Maitland and Newman 2004) and perhaps in tourism world cities we may see a pattern of development that echoes Butler's famous tourism area lifecycle (Butler 1980) in which areas are 'discovered' by a few then eventually develop for mass tourism. This perspective would suggest that as new tourism areas mature and attract more visitors, 'explorers' who want to get off the beaten track would look for new areas, and certainly in London there is evidence of some visitors seeking out newly gentrifying areas like Shoreditch and Hoxton, and moving further afield. Whilst the areas we have examined were unplanned we must consider whether planning policy can intervene to stabilise them if that is thought desirable, or whether they will inevitably evolve into established tourism precincts, or simply lose appeal. Perhaps new areas must continually emerge to meet the changing demands of city users. We need longitudinal research to examine the development of areas over time if we are to gain a better understanding of their changing role and of the interaction of tourism with other development processes in the city.

We defined world tourism cities in terms of their complex economic and social structures and polycentric spatial forms in addition to their well-known historic attractions and iconic buildings. The 'old world' cities have a greater depth of historical foundation but both old and new world city share multifunctional roles that attract and retain talent and both display character and tradition through their spatial diversity and distinctiveness. These cities have substantial tourism histories. In Berlin for example, Potsdammer Platz was one of the first modern shopping destinations, and in the nineteenth and early twentieth centuries the city enjoyed moments of unrivalled reputation for cultural creativity, and New York in the first half of the twentieth century drew visitors to be stunned by the scale and hyperactivity of modernity. These city reputations and images were created out of economic and cultural change, and through changing urban forms. The 'Swinging London' of the 1960s owed something to dynamic culture industries, pioneering gentrification and growing diversity as well as to a crude marketing campaign. The links we emphasise between tourism and other urban processes are not new, but more importantly perhaps not well enough understood within the discrete boundaries of academic literatures. To understand the attractiveness of places off the beaten track in our group of cities we have to look to wider processes of urban change. In many cases it is the conviviality of middle-class residents, workers and visitors, the shared urban preferences of these groups that brings about a (re)discovery of distinctive and desirable urban places. Understanding urban tourism is inseparable from a wider understanding of urban processes. In the places our contributors have examined in these world tourism cities the distinctions between tourists and others and the distinction between tourist places and other spaces is clearly breaking down. How well urban research responds to these challenges will have implications for urban policy. As we have said, in this group of cities new tourism areas have opened up for the most part without the direct intervention of public policy. In the case of East London, government housing policy and local planning policy have encouraged rapid change in housing markets and offered substantial rewards to gentrification. Attracting the middle-class (back) to the City centre is an almost universal policy objective. But when city governments are thinking about housing policy and desirable residents they have not been thinking about or planning for tourism. Visitors have found their way to newly gentrified areas because they also enjoy characterful townscapes and share the consumption preferences of residents. In this group of cities there is awareness of the need to make attractive places, but little evidence that public policy has a systematic understanding of the complex processes of urban transformation that are at work. The policy desire in London, for example, to draw tourists eastwards will not work without better understanding of how visitors have contributed to the transformation of those areas on the fringe of the City where a new urban imagination seems to have taken root.

THE CONTINUING ADVANTAGES OF WORLD TOURISM CITIES

Tourism is forecast to grow to 1.6 billion arrivals worldwide by 2020 – an increase of some 700 million on the 2007 total. The UNWTO believes that this forecast should be maintained despite recent events and uncertainties because the underlying structural trends on which it is based have not changed significantly (UNWTO 2008). If that is so, our world tourism cities can expect many more visitors. Some will be first timers, travelling in organised groups and eager to see the most familiar sights. Accommodating greater numbers of visitors in tourist hotspots is likely to be problematic. These tourism precincts will need deliberate and careful planning and management. However, many of those who come to world tourism cities will be experienced city users, in the city for a variety of reasons beyond simply a holiday, and often with an attachment to the place. For them, finding distinctive places in a globalised world will matter more than new icons or attractions, and they will seek areas not designed as tourism zones, in which their consumption preferences are matched by those of city residents. In these areas, the absence of deliberate tourism planning forms part of the attraction, and the ability of world tourism cities to continue to generate new areas like this gives them a decisive competitive advantage.

The structural trends on which the UNWTO forecast is based may change in future – for example through rising oil prices, serious action to combat climate change or significantly increased terrorism. If the cost, risk and inconvenience of travel increase significantly, then trends will change. However, world tourism cities seem well placed to adapt. Their iconic attractions will continue to appeal to first time visitors, perhaps willing to pay high prices for 'the trip of a lifetime'. However, they will also attract a broader range of visitors and temporary residents who are working, keeping in touch with family, or continuing their education as well as enjoying leisure. For these people, distinctive and convivial areas catering for a diverse range of users will be a vital part of a city's attractions. The ability of world tourism cities to combine icons, tourism precincts and evolving off the beaten track areas gives them a decisive advantage in adapting to change.

Notes

1 Developing world tourism cities

1 *The Times*, 5th January 2008, p. 43. *Guardian*, 6th March 2008, p. 12.

2 New York tourism: dual markets, duel agendas

1 Data for this research is derived from the following sources in which survey interviews were conducted with a total of 220 tourists. The sample was derived from two sources, a pilot study was conducted in 2004 by Roberto Genoves, Mark Hogan, Phillip Mallory, Megan Murphy and Jose Roman – students participating in an applied urban research workshop at the Hunter College Graduate Program in Urban Affairs and Planning. This pilot study (modelled upon Maitland and Newman's New Tourism Areas survey), produced a total of seventy two usable survey interviews with tourists in two study areas (Downtown Brooklyn (n = 32), Long Island City Queens (n = 40). In 2005, funding was received from PSCCuny to expand the study and with the support of a team of hired survey researchers, an additional 147 survey interviews were conducted with tourists in two study areas – Downtown Brooklyn a (n = 72) and Harlem Manhattan (n = 75). Between 1999 and 2008, thirty key informant interviews were conducted with politicians, community development experts, and tourism professionals in Brooklyn, Queens and Manhattan. Finally, in 2000 a study of tourism coalitions was conducted in downtown Brooklyn, a working paper was produced from this, that is used throughout this paper (Gross and Rogowsky 2000).
2 The NYC and Company Web site was checked monthly between September 2007 and May 2008. Counts of listings by borough, in each area mentioned, were conducted during this time period. The percentages identified represent the average across the entire time period.

6 Sydney: beyond iconicity

1 One of the early acts of this new state government, elected in May 1976, was to introduce the Heritage Act 1977, the first legislation in NSW aimed at protecting the state's built heritage. Later, this government enacted new planning legislation, the Environmental Planning and Assessment Act 1979, that substantially increased legitimate avenues for public participation in planning. However, throughout its life SCRA remained exempt from the requirements of this legislation.
2 Many package tours to Australia at this time featured Sydney, the Great Barrier Reef in North Queensland and Ayers Rock – now Uluru – in the Northern Territory, arguably Australia's most iconic tourist sights. These tours were generally of one to two weeks' duration.

3 During the 1980s, the Sydney Cove Redevelopment Authority dropped the word 'Redevelopment' from its title in order to improve its somewhat tarnished image and convey the message that it was no longer concerned with the wholesale redevelopment of one of Australia's most important historic areas.
4 Examining visitation trends over a longer period than this is made difficult by a change in visit estimation methods in 2003. Figures before and after that year are not directly comparable.

Bibliography

Abu-Lughod, J. (1999) *New York, Chicago, Los Angeles: America's Global Cities*, Minneapolis: University of Minnesota Press.

Allaman M. (2002) 'Bercy village, un grand Cru', *Diagonal*, 154: 45–8.

Allmendinger, P. (2001) *Planning in Postmodern Times*, London: Routledge.

Allnut, D. (2004) 'Review of Tourism Statistics', *National Statistics Quality Review*, 33.

ALTAREA (2003) 'Bercy village'. Online. Available at http://www.altarea.com/images/Activities/etudes/pdf/bercy-village.pdf (accessed 25 August 2007).

APUR (1974) 'Paris Sud-Est', *Paris Projet*, 12.

Ashworth, G. J. and Dietvorst, A. G. (1995) *Tourism and Spatial Transformations*, Wallingford: CABI.

Ashworth, G. J. and Tunbridge, J. E. (2000) *The Tourist-Historic City*, 2nd edn, Amsterdam: Pergamon.

Association, «Ca se visite». Online. Available at http://www.ca-se-visite.fr/ (accessed 20 August 2007).

Atkinson, R. and Bridge, G. (2005) 'Introduction', in *Gentrification in a Global Context: The new urban colonialism*, London: Routledge, 1–17.

Baldwin, J. H. (2006) *Who's Next? Questioning the Future of Museum Leadership in New York State*, New York: Museum Association of New York, 20.

Baudoin, P. (2000) 'Rapport relatif au schéma régional de développement du tourisme et des loisirs d'Ile-de-France pour 2000–2010', work paper, Paris: CESR (Conseil Economique et Social Région Ile de France).

Beioley, S., Maitland, R. and Vaughan, R. (1990) 'Tourism and the Inner City', London: HMSO.

Bell, J. (1998) 'Disney's Times Square: The New American Community Theatre', *TDR/The Drama Review*, 42(1), Spring 1998.

Benedictus, L. (2005) 'Every race, colour, nation and religion on earth'. The Guardian, 21 January 2005, www.guardian.co.uk, accessed 28 October 2007.

Bernt, M. (2003) *Rübergeklappt. Die 'Behutsame Stadterneuerung' im Berlin der 90er Jahre*, Berlin: Schiler Verlag.

Bernt, M. and Holm, A. (2002) 'Gentrification in East Germany: the case of Prenzlauer Berg', *Deutsche Zeitschrift für Kommunalwissenschaften*. Online. Available at http://www.difu.de/publikationen/dfk/en/02_2/02_2_bernt.shtml (accessed 6 June 2007).

Blum, A. (2005) *The imaginative structure of the city*, Montreal: McGill – Queens University Press.

Bockmeyer, J. (2006) 'Social Cities and Social Exclusion. Assessing the Role of Turkish Residents in Building the "New Berlin"', *German Politics and Society*, 24 (4): 49–78.

Bouinot, J. (2002) *La Ville Compétitive*, Paris: Economica.

Bram, J. (1995) 'Tourism and New York City's Economy', *Current Issues in Economic and Finance*, 1(7), October 1995.

Bremner, C. (2007) 'Top 150 City Destinations: London Leads the Way', *Euromonitor International*, 11 October 2007.

Bridge, G. (2007) 'A Global Gentrifier Class?', *Environment and Planning A*, 39: 32–46.

Bridge, G. and. Watson, S. (2000) 'City economies', in Bridge, G. and Watson, S., eds, *A Companion to the City*, Oxford: Blackwell.

Brindley, T., Rydin Y. and Stoker, G. (1996) 'Remaking Planning', London: Routledge.

BTA (1998) 'Training for tourism Today and tomorrow', in *Development strategy for London*, London: British Tourist Authority.

BTM (2007a) 'Basistext "Destination Berlin"'. Online. Available at http://www.berlin-tourist-information.de/deutsch/presse/download/d_pr_552_wirtschaft.pdf (accessed 6 June 2007).

BTM (2007b) 'Berlin-Tourismus lässt die Kassen klingeln'. Online. Available at http://www.btm.de/deutsch/presse/download/d_pr_552_wirtschaft.pdf (accessed 6 December 2007).

BTM (2007c) *Berlin registriert weitere Besucherzuwächse im Monat Juli. Bereits 9, 68 Millionen Übernachtungen in den ersten sieben Monaten verzeichnet*, Berlin: Pressemitteilung.

BTM (2007d) 'Berlin – Cultural Metropolis'. Online. Available at http://www.berlin-tourist-information.de/english/kultur/index.php (accessed 6 June 2007).

BTM (2002) *Fast jeder zweite internationale Besucher Berlins ist unter 35 Jahre*, Berlin: Pressemitteilung, 05.09.2002.

Building Design & Construction (2008) 'Time Warner Center Named The Office Building of the Year by the Mid Atlantic Conference of Building Owners and Managers Association; New York City's Newest Legacy Building Celebrates Prestigious Award', 8 May 2008, www.bdcnetwork.com.

Bull, P. and Church, A. (2001) 'Understanding urban tourism: London in the early 1990s', *International Journal of Tourism Research*, 3: 141–50.

Butler, R. (1980) 'The concept of a tourist area cycle of evolution, implications for management of resources', *Canadian Geographer*, 24(1): 5–12.

Butler, T. (2007) 'Gentrification', in Buck, N., Gordon, I., Ahrding, A. and Turok, I., *Changing Cities*. Basingstoke: Palgrave, 172–87.

Butler, T. (2003) 'Living in the bubble: Gentrification and its "others" in north London', *Urban Studies*, 40(12): 2469–86.

Camus, J. S. (2001) 'Bercy Village, premier urban entertainment center à la française', *Espaces Loisirs Tourisme Environnement*, 178: 40–4.

Carpenter, H. (1999) *Islington: The Economic Impact of Visitors*, London: Discover Islington.

Chambre de Commerce et d'industrie de Paris (2007a) 'Tourisme en Ile-de-France. Les défis à relever', *Le nouveau courrier*, 142, Avril/Mai 2007. Online. Available at http://www.lenouveaucourrier.ccip.fr/article.asp?id=926.

Chambre de Commerce et d'industrie de Paris (2007b) 'Tourisme en Ile-de-France. Préparer l'avenir', *Le nouveau courrier*, 142, Avril/Mai 2007. Online. Available at http://www.lenouveaucourrier.ccip.fr/article.asp?id=927.

Chambre de Commerce et d'industrie de Paris (2007c) 'Tourisme en Ile-de-France. Un enjeu économique de taille', *Le nouveau courrier*, 142, Avril/Mai 2007. Online. Available at http://www.lenouveaucourrier.ccip.fr./article.asp?id=925.

Chambre de Commerce et d'industrie de Paris (2006) 'Paris, déjà capitale de tourisme d'affaires, prépare les enjeux de demain', *Communiqué de presse deCCIP*, 16 November 2006. Online. Available at http://www.ccip.fr/upload/pdf/CP_Paris_ deja_capitale_du_tourisme_d_affaires_prepare_les_enjeux_de_demain.pdf.

Clark, T. N. (2003) *The City as an Entertainment Machine*, San Diego: Elsevier.

Clarke, M., 'The need for a more critical approach to dockland renewal', in B. Hoyle, D. A. Pinder and M. S. Hussain, eds, *Revitalizing the waterfront: international dimensions of docklands developments*, London: Belhaven Press, 222–31.

Clendinen, D. (1980) 'City Thrives on Alien Horde', New York Times, 28 September 1980.

Coaffee, J. (2003) *Terrorism, risk and the city*, Aldershot: Ashgate.

Cohen, E. (2004) *Contemporary Tourism: Diversity and Change*, New York: Elsevier.

Cohen, E. (1995) 'Contemporary Tourism – Trends and Challenges: Sustainable authenticity or contrived post-modernity?', in Butler, R. and Pearce, D., *Change in Tourism – People, places, processes*, London: Routledge.

Cohen, E. (1972) 'Towards a Sociology of International Tourism', *Social Research*, 39: 164–82.

Conroux, P. (2007) 'Attractivité et développement des territoires. Le nouvel enjeu du tourisme dans le Val-de-Marne', Attractivité des territoires seminar, Paris 13th February 2007.

Cotton, V. (2003) 'Introduction to The London Leadership Forum on Business Tourism', event held in London on 28th of March 2003, London.

Countryside Commission (2005) *Day Visits Survey 2002–2003*, Cheltenham: Countryside Commission.

Coupland, A., ed. (1997) *Reclaiming the City*, London: Spon.

Craig-Smith, S. J. and Fagence, M., eds, *Recreation and Tourism as a Catalyst for Urban Waterfront Redevelopment: An International Survey*, Westport.

Creswell, T. (2004) *Place: A Short Introduction*, Carlton: Blackwell Publishing.

DAVEZIES, L. (2008) *La circulation invisible des richesses: La république des idées*, Paris: Threshold.

Davidson, M. and Lees, L. (2005) 'New-build "gentrification" and London's riverside renaissance', *Environment and Planning A*, 37: 1165–90.

Dávila, A. (2004) 'Empowered Culture? New York City's Empowerment Zone And The Selling Of El Barrio', *The Annals of the American Academy of Political and Social Science*, 594: 49–64.

Dench, G., Gavron, K. and Young, M. (2006) *The New East End. Kinship, Race and Conflict*, London: Profile Books.

Denzin, N. K. (1989) *Interpretive Interactionism*, Newbury Park, CA: Sage.

Department of Employment (1985) *Pleasure, Leisure and Jobs: the business of tourism*, London: HMSO.

DETR (2000) *Our Towns and Cities: The Future-Delivering an Urban Renaissance*, London: Stationery Office.

DETR (1999) *Towards and Urban Renaissance*, London: Stationery Office.

Diehl, V. S., Sundermeier, J. and Labisch, W. (2002) *Kreuzbergbuch*, Berlin: Verbrecher Verlag.

Discover Islington (1992) *A Tourism Strategy and Programme for Islington*, London.

Eade, J. (2002) 'Adventure Tourists and Locals in a Global City', in Coleman, S. and Crang, M., *Tourism: between Place and Performance*, New York: Beghahn Books.

Eisinger, P. (2000) 'The Politics of Bread and Circuses: Building the city for the visitor class', *Urban Affairs Review*, 35(3), January 2000.

Emporis (2008) 'Skyline Ranking', February 2008. Online. Available at http://www.emporis.com.

ETB (1981) *Tourism and urban regeneration: some lessons from American cities*, London: English Tourist Board.

Fainstein, S. (2007) 'Tourism and the Commodification of Urban Culture', *The Urban Reinventors*, 2, November 2007. Online. Available at http://urbanreinventors.net/fainstein.html (accessed 17 January 2008).

Fainstein, S., Hoffman, L. and Judd, D. (2003) 'Making Theoretical Sense of Tourism', in *Cities and Visitors: Regulating People, Markets and City Space*, Oxford: Blackwell, 239–53.

Fainstein, S. and Gladstone, D. (1999) 'Evaluating Urban Tourism', in Judd, D. R. and Fainstein, S. S., eds, *The Tourist City*, Yale: Yale University Press, 27.

Farias, I. (2008) 'Touring Berlin: Virtual Destination, Tourist Communication and the Multiple City', PhD Dissertation, Berlin: Humboldt University of Berlin.

Feifer, M. (1986) *Tourism in History*, New York: Stein and Day.

Feron, J. (1979) ' "I Love N.Y." New Statewide Affair', New York Times, 25 March 1979.

Firestone, D. (1994) 'For Cultural Institutions, Art of Collaboration Is Biggest Show in Town', New York Times, 18 July 1994.

Florida, R. (2004) *The Rise of the Creative Class*, New York: Basic Books.

Franklin, A. and Crang, M. (2001) 'The Trouble with Tourism and Travel Theory?', *Tourist Studies* 1(1): 5–22.

Franklin, A. (2003) *Tourism. An Introduction*, London: Sage.

French Ministry of tourism (2007) 'Bilan de l'année 2006', 7 February 2007. Online. Available at http://www.tourisme.gouv.fr/fr/z2/stat/bilans/att00014546/bilan_touristique2006.pdf (accessed 25 July 2007).

French Ministry of tourism, Tourism Directorate (2006a) 'Key facts on Tourism 2006 edition'. Online. Available at http://www.tourisme.gouv.fr/fr/z2/stat/chiffres/att00009212/ChiffresCles2005_anglais.pdf (accessed 25 July 2007).

French Ministry of tourism, Tourism Directorate (2006b) 'Poids économique du tourisme (en 2005)'. Online. Available at http://www.tourisme.equipement.gouv.fr/fr/navd/dossiers/taz/att00002082/poids_economique07.pdf (accessed 23 July 2007).

French Ministry of tourism, Tourism Directorate (2006c) 'Politique du tourism (for the year of 2007)'. Online. Available at http://www.tourisme.equipement.gouv.fr/fr/navd/dossiers/taz/att00002082/politique_tourisme07.pdf (accessed 19 July 2007).

Genoves, R., Hogan, M., Mallory, P. and Roman, J. (2004) 'The Real New York', unpublished study on behalf of the Centre for Tourism, London: University of Westminster.

Getz, D. (1994) 'Event Tourism and the Authenticity Dilemma', in Theobold, W. F., *Global Tourism*, Oxford: Butterworth Heinemann.

Gilbert, D. and Hancock, C. (2006) 'New York City and the Transatlantic Imagination: French and English Tourism and the Spectacle of the Modern Metropolis, 1893–1939', *Journal Of Urban History*, 33(1): 77–107.

GLA Economics (2003) *Spending Time: London's Leisure Economy*, London: Greater London Authority.

Gladstone, D., and Fainstein, S. (2001) 'Tourism in US Global Cities: A Comparison of New York and Los Angeles', *Journal of Urban Affairs*, 23(1): 23–40.

Glaeser, E., Kolko, L. And Saiz, I. (2000) *Consumer City*, Harvard: Harvard Institute of Economic Research.

Graham, S. (2004) 'Introduction: cities warfare, and states of emergency', in *Cities War and Terrorism*, Oxford: Blackwell, 1–25.

Gotham, F. K. (2005) 'Tourism Gentrification: The Case Of New Orleans' Vieux Carre (French Quarter)', *Urban Studies*, 42(7): 1099–1121.

Gottdiener, M. (2001) *The theming of America*, Boulder, CO: Westview Press.

Greenberg, M. (2006) 'Urban restructuring and tourism marketing: the dual transformation of neoliberal New York', ASA conference, Montreal.

Griffin, T. (1989) 'Hotel development: the case of downtown Sydney', in Blackwell, J. and Stear, L., eds, *Case Histories of Tourism & Hospitality*, Sydney: Australian-International Magazine Services, 317–33.

Griffin, T. and Darcy, S. (1997) 'Australia: consequences of the newly adopted pro-Asia orientation', in Go, F. M. and Jenkins, C. L., eds, *Tourism and Economic Development in Asia and Australasia*, London: Cassell, 67–90.

Griffin, T., Hayllar, B., and King, B. (2006) 'City spaces, tourist places? An examination of tourist experiences in Melbourne's riverside precincts', in O'Mahony, G. B. and Whitelaw, P. A., eds, *CAUTHE: Proceedings of 16th Annual Conference: To the City and Beyond*, CD-ROM. Victoria University, Melbourne, 6–9 February 2006, 1036–50.

Gross, J. S. (2005) 'Cyber Cities', in Wagner, F. W., Mumphry, A., Joder, T. and Akundi, K., eds, *Revitalizing the City: Strategies to Contain Sprawl and Revive the Core*, Armonk, NY: M. E. Sharpe.

Gross, J. S. and Rogowsky, E. T. (2000) 'Tourism and Economic Development: Coalition and Social Capital in Brooklyn', a working paper, The National Center for the Revitalization of Central Cities, College of Urban and Public Affairs, University of New Orleans, November 2000.

Haberman, C. (2003) ' "NYC" Not Poifect, Dem Movies of Brooklyn', New York Times, 31 October 2003.

Hackworth, J. and Rekers, J. (2005) 'Ethnic Packagaing and Gentrification', *Urban Affairs Review*, 42(2): 211–36.

Hall, C. M. (2007) 'Response to Yeoman et al: The fakery of "The authentic tourist" ', *Tourism Management*, 28(4): 1139–40.

Hall, C. M. (1998) 'The politics of decision-making and top-down planning: Darling Harbour, Sydney', in Tyler, D., Guerrier, Y. and Robertson, M., eds, *Managing Tourism in Cities: Policy, Process and Practice*. Chichester: John Wiley & Sons, 9–24.

Harpaz, B. J. (2006) 'Brooklyn: A destination on its own', The Buffalo News, 10 September 2006.

Harvey, D. (1989) *The Condition of Postmodernity*, Oxford: Blackwell.

Harvey, D. C. and Lorenzen, J. (2006) 'Signifying Practices and the Co-Tourist', *Tourismos*, 1(1): 9–26.

Hatem, F. (2004) *Investissement international et politique d'attractivité*, Paris: Economica.

Häussermann, H. and Colomb, C. (2003) 'The New Berlin: Marketing the City of Dreams', in Hoffman, L. M., Fainstein, S. S., Judd, D. and, Dennis, R., eds,

Cities and Visitors. Regulating People, Markets, and City Space, Malden, MA: Blackwell, 200–18.

Häussermann, H. and Kapphan, A. (2000) *Berlin – Von Der Geteilten Zur Gespaltenen Stadt? Sozialräumlicher Wandel seit 1990*, Opladen: Leske & Budrich.

Hayllar, B. and Griffin, T. (2007) 'A tale of two precincts', in Tribe, J. and Airey, D., eds, *Developments in Tourism Research*, Oxford: Elsevier, 155–69.

Hayllar, B. and Griffin, T. (2005) 'The precinct experience: a phenomenological approach', *Tourism Management*, 26(4): 517–28.

Hayllar, B., Griffin, T. and Edwards, D., eds. (2008) *City Places: Tourist Spaces*, London: Elsevier.

Herden Studienreisen (2006) *Berlin for young people*, English Edition, Berlin.

Hoffman, L. M. (2003a) 'The Marketing of Diversity in the Inner City: Tourism and Regulation in Harlem', *International Journal of Urban and Regional Research*, 27(2), June 2003: 286–99.

Hoffman, L. M. (2003b) 'Revalorizing the Inner City: Tourism and Regulation in Harlem', in Hoffman, L. M., Fainstein, S. S. and Judd, D. R., eds, *Cities and Visitors*. Malden, MA: Blackwell Publishing.

Hopper, P. (2003) 'Marketing London in a Difficult Climate', *Journal of Vacation Marketing*, 9(1): 81–8.

Hotel & Tourism Asset Advisory Services and JLW Transact (1994) *Sydney and Environs Accommodation Demand and Supply Study: 1994–2000*, Sydney: Toursim Olympic Forum.

Huning, S. and Novy, J. (2006) 'Tourism as an Engine of Neighborhood Regeneration? Some Remarks Towards a Better Understanding of Urban tourism beyond the "Beaten Path"', CMS Working Paper Series 006-2006, Berlin: Centre for Metropolitan Studies.

Huxley, M. (1991) 'Darling Harbour and the immobilisation of the spectacle', in Carroll, P., Donohue, K., McGovern, M. and McMillen, J., eds, *Tourism in Australia*, Sydney: Harcourt Brace Jovanovich, 141–52.

Huyssen, A. (2003) *Present pasts: urban palimpsests and the politics of memory*, Stanford: Stanford UP.

IAURIF, Institut d'Aménagement et d'Urbanisme de la Region d'Ile de France (2006) 'Les pôles touristiques Régionaux prioritaires en Ile de France Mars', working paper, IAURIF, Paris, March 2006.

IAURIF, Institut d'Aménagement et d'Urbanisme de la Region d'Ile de France (2006) 'Note rapide: Population-Modes de vie', March 2006. Online. Available at http://www.iaurif.org/fr/ressources_doc/publications/publicationsrecentes/notesrapides/pdf/pop_mod_vie/nr_414.pdf (accessed 6 March 2008).

Ingallina, P. and Alcaud, D., 'L'attractivité des territoires', in Alcaud, D., *Manuel de culture générale en aménagement*, Paris 2008.

Ingallina, P. and Park, J. Y. (2005) 'City marketing et espaces de consommation: Les nouveaux enjeux de l'attractivité urbaine', *Urbanisme*, 344: 64–7.

Ingallina P., *Il Progetto urbano. Dall'esperienza francese alla realtà italiana*, Milan 2004.

Ingallina, P. (2001) *Le Projet Urbain*, Paris: Vendôme.

INSEE (2005) *Le tourisme en France: édition 2005*, Paris: INSEE.

Jacobs, J. (1961) *The Death and Life of Great American Cities*, New York: Random House.

Judd, D. (2003) 'Visitors and the Spatial Ecology of the City', in Hoffman, L. and Judd, F. S. S. and D., *Cities and Visitors*, Oxford: Blackwell, 23–38.

Judd, D. (1998) 'Constructing the Tourism Bubble', in Judd, D. and Kantor, P., eds, *The Politics of Urban America: A Reader*, 2nd Edition, Needham Heights, MA: Allyn and Bacon.

Kil, W. and Silver, H. (2006) 'From Kreuzberg to Marzahn. New Migrant Communities in Berlin', *German Politics and Society*, 24(4): 95–121.

King, A. (2006) 'World Cities: Global? Postcolonial? Postimperial? or just the result of happenstance? Some cultural comments', in Brenner, N. and Keil, R., *The Global Cities Reader*. London: Routledge, 319–24.

Kong, L. (2007) 'Cultural icons and urban development in Asia: Economic imperative, national identity, and global city status', *Political Geography*, 26(4): 383–404.

Krajewski, C. (2005) 'Städte-Tourismus im "Neuen Berlin"', in Landgrebe, S. and Schnell, P., eds, *Städtetourismus*, Munich: Oldenbourg Wissenschaftsverlag, 279–95.

Krätke, S. (2004) 'City of Talents? Berlin's Regional Economy, Socio-Spatial Fabric and "Worst Practice" Urban Governance', *International Journal of Urban and Regional Research*, 28(3): 511–29.

Krätke, S. (2003) 'Global media cities in a world-wide urban network', *European Planning Studies*, 11(6): 605–28.

Kuntzman, G. (2004) 'YO, CHECK US OUT: Pitching outer boros to tourists', New York Post, 12 February 2004.

Lang, B. (1998) *Mythos Kreuzberg. Ethnographie eines Stadtteils 1961–1995*, Frankfurt and New York:Campus.

Lees, L. (2007) 'Progress in gentrification research?', *Environment and Planning A*, 39: 228–34.

Lefebvre, H. (1991) *The Production of Space*, Oxford: Blackwell Publishing.

Lehrer, U. (2006) 'Willing the Global City: Berlin's Cultural Strategies of Interurban Competition After 1989', in Brenner, N. and Keil, R., eds, *The Global City Reader*, London: Routledge, 332–38.

Lichtman, F. (2006) 'BRIC: Arts Umbrella Organization Is Vital Part of Brooklyn'. Brooklyn Daily Eagle, 17 May 2006, www.brooklyneagle.com.

Lipscomb, D. and Weatheritt, L. (1977) 'Some economic aspects of tourism in London', *Greater London Intelligence Journal*, 42: 15–17.

Lloyd, R. (2005) *Neo-bohemia:art and commerce in the post-industrial city*, New York: Routledge.

Lloyd, R. (2002) 'New Bohemia. Art and Neighborhood Redevelopment in Chicago', *Journal of Urban Affairs*, 24(5): 517–32.

Lloyd, R. (2000) Grit as Glamour, cited in Judd, D., 2003, Visitors and the Spatial Ecology of the City, in L. Hoffman, S. Fainstein, D. Judd, eds, Cities and Visitors, Oxford: Blackwell, 23–38.

London Councils (2006) 'London Tourism: by numbers', *London Bulletin*, July/August 2006.

London Development Agency (2006) *London Tourism Action Plan 2006–2009*, London: London Development Agency.

London Development Agency (2004) *London Tourism Action Plan 2003/4–2005/6*, London: London Development Agency.

London Development Agency (2002) *London Tourism Action Plan 2003/4–2005/6*, London: London Development Agency.

London ICC (2005a) *ICC Commission Report*, London: London ICC.

London ICC (2005b) *Commission Report – Initial Consultation Exercise*, London: London ICC.

London Tourist Board and Convention Bureau (1998) *London Tourism Statistics 1998*, London: London Tourist Board and Convention Bureau.

London Tourist Board and Convention Bureau (1997) *Tourism Strategy for London 1997–2000*, London: London Tourist Board and Convention Bureau.

Long, P. (1999) 'Tourism Development Regimes in the Inner City fringe: the case of discover Islington, London', *Journal of Sustainable Tourism*, 8(3): 190–206.

Lynch, K. (1960) *The Image of the City*, Cambridge: MIT Press.

MacCannell, D. (1999) *The Tourist: A New Theory of the Leisure Class*, Third Edition, Berkeley: University of California Press.

MacCannell, D. (1976) *The Tourist: A New Theory of the Leisure Class*, New York: Sulouker Books.

Maguire, J. (2005) *Power and Global Sport: Zones of prestige, emulation and resistance*, London: Routledge.

Maitland, R. (2008) 'Conviviality and Everyday Life: the Appeal of New Areas of London for Visitors,' *International Journal of Tourism Research*, 10(1): 15–25.

Maitland, R. (2007a) 'Cultural Tourism and the Development of New Tourism Areas in London', in Richards, G., *Cultural Tourism – Global and Local Perspectives*, Binghampton, New York: Haworth Press, 113–30.

Maitland, R. (2007b) 'Tourists, the creative class, and distinctive areas in major cities', in Richards, G. and Wilson, J., *Tourism, Creativity and Development*, London and New York: Routledge.

Maitland, R. (2006) 'Culture, City Users and the creation of new tourism areas in cities', in Smith, M., *Tourism, Culture and Regeneration*, London: CABI.

Maitland, R. (2003) 'Cultural Tourism and new tourism areas', in *Cultural tourism: globalising the local – localising the global*, Barcelona: ATLAS. Tilburg.

Maitland, R. and Newman, P. (2004) 'Developing Metropolitan Tourism on the Fringe of Central London', *International Journal of Tourism Research*, 6: 339–48.

Marcuse, P. (2006) 'The Down side dangers in the Social City Program', *German Politics and Society*, 24(4): 122–30.

Martin, G. P. (2005) 'Narratives Great and Small: Neighbourhood Change, Place and Identity in Notting Hill', *International Journal of Urban and Regional Research*, 29(1): 67–88.

Martinotti, G. (1999) 'A City For Whom? Transients And Public Life In The Second-Generation Metropolis', in Beauregard, R. and Body-Gendrot, S., *In The Urban Moment*, London: Sage, 155–84.

May, J. (1996a) 'In search of authenticity off and *on* the beaten track', *Environment and Planning D: Society and Space*, 14: 709–36.

May, J. (1996b) 'Globalization and the politics of place: place and identity in an inner London neighbourhood', *Transactions of the Institute of British Geographers*, 21(1): 194–215.

Mayer, M. (2006a) 'New Lines of Division in the New Berlin', in Ulfers, F., Lenz, G. and Dallmann, A., eds, *Towards a New Metropolitanism: Reconstituting Public Culture, Urban Citizenship, and the Multicultural Imaginary in New York City and Berlin*, Heidelberg: Universitätsverlag Winter, 171–83.

Mayer, M. (2006b) 'Berlin Nonprofits in the reshaping of Welfare and Employment Policies', *German Politics and Society*, 24(4): 131–44.

Mayor of London (2004a) 'London Cultural Capital', London: GLA.

Mayor of London (2004b) *The London Plan: Spatial Development Strategy for Greater London*, London: GLA.

McCabe, S. (2005) 'Who is a Tourist? A critical review', *Tourist Studies*, 5(1): 85–106.

McCarthy, J. (1996) 'The evolution of planning approaches: North Southwark 1971–1994', *Land Use Policy*, 13(2): 149–51.

McKercher, B. (2002) 'Towards a Classification of Cultural Tourists', *International Journal of Tourism Research*, 4: 29–38.

McKinsey & Co. (2002) *Cultural Capital: Investing in New York's Economic and Social Health*, New York: McKinsey & Co.

Meethan, K. (2001) *Tourism in a Global Society*, Basingstoke: Palgrave.

Mehta, J. (2004) 'Long Island City becomes and unlikely arts enclave', *City beat: Long Island City*, March 2004.

Mele, C. (2000) *Selling the Lower East Side: Culture, Real Estate, and Resistance in New York City*, Minneapolis: University of Minnesota Press.

Michael, E. J., (2008) *Tourism in Australia – where have all the people gone? Competition in Tourism: Business and Destination Perspectives*, Helsinki: Travel and Tourism Research Association.

Minnite, L. i. J. e. (2005) 'Outside the Circle: the impact of post 9/11 responses on immigrant communities in New York', in Mollenkopf, J., *Contentious City. The Politics of recovery in New York City*, New York: Russell Sage Foundation, 165–204.

Montgomery, J. (2004) 'Cultural quarters as mechanisms for urban regeneration. Part 2: a review of four cultural quarters in the UK, Ireland and Australia', *Planning, Practice and Research*, 19(1): 3–31.

Moore, F. (2004) 'Tales of the global city: German expatriate employees, globalisation and social mapping', *Anthropology Matters Journal*, 6(1): 12.

Munro, C. (2007) 'Welcome to the CBD: all arteries no pulse', Sydney Morning Herald, 12 November 2007: 1.

Neill, W. (2004) *Urban Planning and Cultural Identity*, London: Routledge.

New York City Department of City Planning (2006a) *Borough of Brooklyn*, December 2006. Online. Available at http://www.nyc.gov/html/lucds/cdstart.shtml.

New York City Department of City Planning (2006b) *Brooklyn Community District 2*, December 2006. Online. Available at http://www.nyc.gov/html/dcp/html/lucds/cdstart.shtml.

New York City Department of City Planning (2006c) *Brooklyn Community District 8*, December 2006. Online. Available at http://www.nyc.gov/html/dcp/html/lucds/cdstart.shtml.

New York City Department of City Planning (2006d) *Borough of Queens*, December 2006. Online. Available at http://www.nyc.gov/html/dcp/html/lucds/cdstart.shtml.

New York City Department of City Planning (2006e) *Queens Community District 2*, December 2006. Online. Available at http://www.nyc.gov/html/dcp/html/lucds/cdstart.shtml.

New York City Department of City Planning (2006f) *Manhattan Community District 10*, December 2006. Online. Available at http://www.nyc.gov/html/dcp/html/lucds/cdstart.shtml.

New York City Department of City Planning (2006g) *Manhattan Community District 11*, December 2006. Online. Available at http://www.nyc.gov/html/dcp/html/lucds/cdstart.shtml.

New York City Mayor's Office of Industrial and Manufacturing Business (2006) 'Mayor Bloomberg Outlines Latest Successes Of The Administration's Five Borough Economic Development Strategy During Weekly Radio Address', Press Release # 79, 19 March 2006.

New York City OASIS (2008) 'OASIS Map New York City'. Online. Available at http://www.oasisnyc.net (accessed March 2008).

New York City Office of Management and Budget (2007) 'Adopted Budget Fiscal Year 2007'. Online. Available at http://www.home2.nyc.gov/html/omb/pdf/erc6_07.pdf.

New York State Office of the State Comptroller (2008) 'Review of the Financial Plan of the City of New York', Report 10-2008: 3, www.osc.state.ny.us.

Newman, P. and Smith, I. (2000) 'Cultural production, place and politics on the south bank of the Thames', *International Journal of Urban and Regional Research*, 24(1): 9–24.

Nixon, S. (2008) 'The cars that ate Sydney', Sydney Morning Herald, 2 January 2008: 21.

NYC & Company, 'NYC Statisitics', www.nycvisit.com, accessed May 2008.

NYC & Company, 'Visitors to Lower Manhattan & Upper Manhattan', a research study prepared for NYC & Company by Audience Research & Analysis, New York, February 2003.

ONS (2007a) *Focus on London*, London: Office for National Statistics.

ONS (2007b) *Summary of Social Trends 2007*, London: Office for National Statistics.

Page, S. J. (2002) 'Urban tourism: evaluating tourists' experience of urban places', in Ryan, C., *The Tourist Experience*, London: Continuum.

Park, J. (2005a) 'Role of the culture and tourism development in the strategic planning for city attractiveness', congress paper, AESOP Congress, Vienna 2005.

Park, J. (2005b) 'Comprehension of urban consumption spaces in strategies for urban attractiveness improvement', *Italian Journal Regional Science*, 4(3): 69–93.

Paterson, A. B. (1906) 'An Outback Marriage: the Story of Australian Life', in *Push Society*, Chapter 4. Online. Available at http://www.gutenberg.org.

Peel, V., and Steen, A. (2007) 'Victims, hooligans and cash-cows: media representations of the international backpacker in Australia', *Tourism Management*, 28(4): 1057–67.

Pine II, B. J. and Gilmore, J. H. (1999) *The Experience Economy*, Boston: Harvard Business School Press.

Poon, A. (1993) *Tourism, Technology and Competitive Strategies*, Wallingford: CAB International.

Porter, R. (2000) *London: A Social History*, London: Penguin.

Reichl, A. J. (1999) *Reconstructing Times Square: politics and culture in urban development*, Lawrence, Kan.: University Press of Kansas.

Relph, E. (1976) *Place and Placelessness*, London: Pion Books.

Robinson, D. (2004) 'Promoting Sustainable Urban Tourism as an Economic Development Strategy for Queens: What Works', unpublished research paper, New York.

Roddewig, R. J. (1978) *Green Bans: the Birth of Australian Environmental Politics*, Sydney: Hale & Iremonger.

Roder, B. and Tacke, B., eds (2004) *Prenzlauer Berg im Wandel der Geschichte*, Berlin: be.bra verlag.

Rofe, M. (2003) ' "I want to be global": Theorising the gentrifying class as an emergent elite global community', *Urban Studies*, 40(12): 2511–26.

Rojek, C. and Urry, J. (1997) 'Transformations of travel and theory', in *Touring Cultures: transformations of travel and theory*, London and New York: Routledge.

Roncayolo, M. (2003) 'La ville est toujours la ville de quelqu'un', in Lévy J., Mongin O., Paquot T., Roncayolo M. and Cardinali P., *De la ville et du citadin. Savoirs à l'oeuvre*, Marseille: Parentheses.

Rowe, D. and Stevenson, D. (1994) ' "Provincial Paradise": urban tourism and city imaging outside the metropolis', *Australian and New Zealand Journal of Sociology*, 30(2): 178–93.

Sallet-Lavorel, H. (2003) 'Encourager le rapprochement entre visiteurs et franciliens', *Cahier espaces*, 78: 118–33.

Sassen, S. (1991) *The Global City*, Princeton: Princeton University Press.

Savitch, H. and Ardashev, G. (2001) 'Does terror have an urban future?', *Urban Studies*, 38(13): 2515–33.

Scheffler, K. (1989[1910]) *Berlin – Ein Stadtschicksal*, Berlin: Fannei & Walz.

Schulte-Peevers, A. (2002) *Lonely Planet Berlin*, 3rd Edition. Melbourne, Vic. and London: Lonely Planet.

Searle, G. (2008) 'Conflicts and politics in precinct development', in Hayllar, B., Griffin, T. and Edwards, D., eds, *City Places: Tourist Spaces*, London: Elsevier, 203–22.

Selby, M. (2004a) 'Consuming the city: conceptualizing and researching urban tourist knowledge', *Tourism Geographies*, 6(2): 186–207.

Selby, M. (2004b) *Understanding Urban Tourism*, London: I.B. Taurus.

Senatsverwaltung für Stadtentwicklung (2006) 'Orte der Internationalität'. Online. Available at http://www.stadtentwicklung.berlin.de/planen/basisdaten_stadtentwicklung/internationalitaet/ (accessed 6 June 2007).

Senatsverwaltung für Wirtschaft (2007) 'Technologie und Frauen, Wirtschafts- und Arbeitsmarkt-bericht'. Online. Available at http://www.berlin.de/imperia/md/content/senatsverwaltungen/senwaf/wirtschaft/berichte/wab2007.pdf (accessed 6 June 2007).

Seyfried, V. and Peterson, J., 'Historical Essay: A Thumbnail View'. Online. Available at http://www.queensbp.org/content_web/tourism/tourism_history.shtml (accessed 1 March 2008).

Shaw, S., Bagwell, S., and Karmowska, J. (2004) 'Ethnoscapes As Spectacle. Reimaging Multicultural Districts As New Destinations For Leisure And Tourism Consumption', *Urban Studies*, 41(10): 1983–2000.

Sheller, M. and Urry, J. (2006) 'The new mobilities paradigm', *Environment and Planning A*, 38: 207–26.

Sheller, M. and Urry, J. (2004a) 'Places to play, places in play', in *Tourism Mobilities: places to play, places in play*, London: Routledge.

Sheller, M. and Urry, J. (2004b) *Tourism Mobilities: places to play, places in play*, London: Routledge.

Shields, R. (1991) *Places on the Margins*, London and New York: Routledge.

Short, J. R., *The Urban Order: an introduction to cities, culture and power*, Oxford: Blackwell.

Short, J. R., Breitbach, C., Buckman, S. and Essex, J. (2000) 'From world cities to gateway cities', *City*, 4(3): 317–40.

Silbert, J. (2005) 'SOAPBOX; Of Icexream, and Homoginzation', New York Times, 13 March 2005.

Smith, A. (2006) 'Sport events, tourism and urban regeneration', in Smith, M. K., *Tourism, Culture and Regeneration*, Wallingford: CABI.

Souter, T. (1997) 'The New Heyday of Harlem', The Independent, 8 June 1997.

Soysal, L. (2006) 'World City Berlin and the Spectacles of Identity: Public Events, Immigrants and the Politics of performance', MiReKoc – Migration research Program, Koç University. Istanbul.

Statistisches Landesamt Berlin-Brandenburg (2007a) 'Basisdaten Erwerbstätigen-rechnung'. Online. Available at http://www.statistik-berlin-brandenburg.de/ (accessed 1 September 2007).

Statistisches Landesamt Berlin-Brandenburg (2007b) 'Commissioned Data Set'. Online. Available at http://www.statistik-berlin-brandenburg.de/ (accessed 1 September 2007).

Suvantola, J. (2002) *Tourist's Experience of Place*, Aldershot: Ashgate.

Swarns, R. L. (1996) 'Tourists in Search of the "Real" New York', New York Times, 8 July 1996.

Sydney Olympic Park Authority (2008) *Sydney Olympic Park Authority's History*. Online. Available at http://www.sydneyolympicpark.com.au/corporate/history (accessed 13 March 2008).

Talbot, D. (2004) 'Regulation and Racial Differentiation in the Construction of Night-time Economies: A London Case Study', *Urban Studies*, 41(4).

Teedon, P. (2001) 'Designing a Place Called Bankside: On Defining an Unknown Space in London', *European Planning Studies*, 9(4): 459–81.

The Broadway League, 'Broadway Season Statistics', www.livebroadway.com.

The Economist (2006) 'Poor but sexy: Berlin', *The Economist*, 380(8496): 43.

Tourism New South Wales (2007a) Brand Sydney. Online. Available at http://corporate.tourism.nsw.gov.au/default.aspx?PageID=736 (accessed 28 September 2007).

Tourism New South Wales (2007b) Travel to Sydney: Year Ended June 2007. Online. Available at http://corporate.tourism.nsw.gov.au/Sites/SiteID6/objLib12/Sydney%20YE%20Jun%2007.pdf (accessed 28 September 2007).

Tourism New South Wales (2006) *Sydney Surrounds: Summary of Research Findings*, Sydney: TNSW, May 2006.

Tourism New South Wales (1998) *Sydney Tourism Experience Development (STED) Program 1998–1999*, Sydney: TNSW.

Tourism New South Wales (1996) *Attractions for the Future: Attractions Development Strategy for Greater Sydney*, Sydney: TNSW.

Tourism Research Australia (2008a) *International Visitors in Australia – December 2007 Quarterly Results of the International Visitor Survey*, Canberra: Tourism Research Australia.

Tourism Research Australia (2008b) *Travel by Australians – December 2007 Quarterly Results of the National Visitor Survey*, Canberra: Tourism Research Australia.

Tourism Research Australia (2007a) *International Visitors in Australia – June 2007 Quarterly Results of the International Visitor Survey*, Canberra: Tourism Research Australia.

Tourism Research Australia (2007b) *Travel by Australians – June 2007 Quarterly Results of the National Visitor Survey*, Canberra: Tourism Research Australia.

Tran, M. (2005) 'London Celebrates Olympic decision', The Guardian, 6 July 2005.

Treichel, T. and Schmid, E. D. (2005) 'London – Paris – Berlin. Tourismus-Bilanz 2005: Rekordzahl bei Übernachtungen', Berliner Zeitung, 16 December 2005, www.berlinonline.de, accessed 25 January 2007.

TTF Australia (2007a) *Australia's not ready to meet tourism forecasts*, Media Release, 23 November 2007.

TTF Australia (2007b) *Cities crucial to Australian tourism*, Media Release, 14 December 2007.

Tuan, Y. F. (1977) *Space and Place: the Perspective of Experience*, Minneapolis: University of Minnesota Press.

Tyler, D. (1998) 'Getting tourism on the agenda: policy development in the London Borough of Southwark', in Tyler, D., Guerrier, Y. and Robertson, M., *Managing Tourism in Cities: policy, processes and practice*, Chichester: John Wiley & Sons.

U.S. Department of Commerce International Trade Administration, Office of Travel and Tourism Industries (2008) *Overseas Visitation Estimates for U.S. States, Cities, and Census Regions: 2007*, Washington, http://tinet.ita.doc.gov.

UNESCO (2006) 'Berlin City of Design Official Press Release'. Online. Available at http://portal.unesco.org/culture (accessed 12 November 2006).

UNWTO, *Tourism 2020 Vision*, Madrid: United Nations World Tourism Organisation.

Urry, J., ed. (1997) *Touring cultures: transformations of travel and theory*, London and New York: Routledge.

Urry, J. (1995) *Consuming Places*, London: Sage.

Urry, J. (1990) *The Tourist Gaze*, London: Sage.

Van Manen, M. (1990) *Researching Lived Experience*, London and Ontario: State University of New York Press.

Venturi, R. C., Brown, D. S. and Izenour, S. (1972) *Learning from Las Vegas: the Forgotten Symbolism of Architectural Form*, Cambridge, Mass.: MIT Press.

VisitLondon (2007) *London Visitor Statistics 2006/07*, London: VisitLondon.

VisitLondon (2005a) *London for Londoners Part 2: A practical guide for London Boroughs on how to promote tourism to 'locals'*, London: VisitLondon.

VisitLondon (2005b) *London Overseas Visitor Survey 2003/04*, London: VisitLondon.

VisitLondon (2004) *London Visitor Statistics 2003/04*, London: VisitLondon.

Vivant, E. (2007) 'Towards a non-human anthropology of tourism', Paper at the meeting of the Association of Social Anthropologists, London April 2007.

Wang, N. (1999) 'Rethinking authenticity in tourism experience', *Annals of Tourism Research*, 26(2): 349–70.

Ward, J. (2004) 'Berlin, the Virtual Global City', *Journal of Visual Culture*, 3(2): 239–56.

White, E. B. (1949) *Here is New York*, New York: The Little Book Room.

Wickens, E. (2002) 'The sacred and the profane: A Tourist Typology', *Annals of Tourism Research*, 29(3): 834–51.

Williams, A. and Shaw, G. (1998) 'Tourism and the Environment: Sustainability and Economic Restructuring?', in Hall, P. and Lew, A. A., *Sustainable Tourism: A Geographical Perspective*, New York: Longman.

World Tourism Organisation and United Nations (1994) 'Recommendations on Tourism Statistics', Madrid WTO.

Yeoh, B. (2005) 'The Global Cultural City? Spatial Imagineering and Politics in the (Multi)cultural Marketplaces of South-east Asia', *Urban Studies*, 42(5/6).

Yeoman, I., Brass, D. and McMahon-Beattie, U. (2007) 'Current issue in tourism: The authentic tourist', *Tourism Management*, 28(4): 1128–38.

Zandt, D. (2005) 'The Starbucks Tipping Point'. Online. Available at http://www.alternet.org/blogs/themix/23938 (accessed 1 March 2007).

Zukin, S. (2005) *Point of Purchase*, New York: Routledge.

Zukin, S. (1995) *The Cultures of Cities*, Oxford: Blackwell.

Zukin, S. (1991) *Landscapes of Power: From Detroit to Disney World*, Berkeley: University of California Press.

Index